IEG WORLD BANK | IFC | MIGA
INDEPENDENT EVALUATION GROUP

World Bank Group Support
for Innovation and Entrepreneurship

AN INDEPENDENT EVALUATION

ISBN (paper): 978-1-4648-0136-5
ISBN (electronic): 978-1-4648-0137-2
DOI: 10.1596/978-1-4648-0136-5

Cover photo: ©Sergey Nivens. Used with permission; further permission required for reuse.
Cover design: Crabtree + Company

Library of Congress Cataloging-in-Publication Data
World Bank group support for innovation and entrepreneurship : an independent evaluation / World Bank.
      pages cm
   Includes bibliographical references and index.
   ISBN 978-1-4648-0136-5 (alk. paper)
   1. Technological innovations—Economic aspects.
   2. Entrepreneurship. 3. Project management.
I. World Bank.
   HC79.T4W644 2014
   338'.04—dc23
                                      2014015555

Contents

Figures

Tables

Appendices

Abbreviations

AAA	analytic and advisory activity	M&E	monitoring and evaluation
ARD	Agriculture and Rural Development Network	MIC	middle-income country
CAS	Country Assistance Strategy	MIGA	Multilateral Investment Guarantee Agency
CRG	competitive research grants	OECD	Organisation for Economic Co-operation and Development
EIF	Enterprise Incubator Fund		
ESW	economic and sector work	PER	Project Evaluation Report
FDI	foreign direct investment	PSD	private sector development
FPD	Finance and Private Sector Development Network	R&D	research and development
		S&T	science and technology
ICR	Implementation Completion and Results Report	SME	small and medium-size enterprise
ICT	information and communications technology	SPREAD	Sponsored Research and Development Program
IEG	Independent Evaluation Group	STI	science, technology, and innovation
IFC	International Finance Corporation	UNCTAD	United Nations Conference on Trade and Development
IPP	Innovation Policy Platform	XPSR	Expanded Project Supervision Report
IPR	intellectual property rights		
ITE	Innovation, Technology, and Entrepreneurship Practice		

Acknowledgments

This evaluation of the World Bank Group's support for innovation and entrepreneurship was prepared by an Independent Evaluation Group (IEG) team led by Ade Freeman. It was carried out under the guidance of Marvin Taylor-Dormond (Director, Private Sector Evaluation) and Stoyan Tenev (Manager, Private Sector Evaluation) and the direction of Caroline Heider (Director General, Evaluation).

Team members (in alphabetical order) were Unurjargal Demberel, Hiroyuki Hatashima, Houqi Hong, Giuseppe Iarossi, Edna Massay Kallon, Maria Kopyta, Brett J. Libresco, Sara R. Mareno, Ketevan Nozadze, Srinath Sinha, Bowen Patrick Uhlenkamp, Andrew Warner, and Izlem Yenice. Carl J. Dahlman prepared a background paper for the study. IEG colleagues provided valuable feedback at several stages of the evaluation.

The peer reviewers were Natalia Agapitova (World Bank Institute), Philip Auerswald (Associate Professor, George Mason University), and Charles Wessner (Director, Program on Technology, Innovation, and Entrepreneurship, National Academy of Sciences).

Heather Dittbrenner and William Hurlbut edited the report. Emelda T. Cudilla provided administrative support and helped with formatting of the report.

Overview | SUMMARY

Growth in any economy comes from (i) growth in inputs of production,
(ii) improvements in the efficiency of allocation of inputs across economic activities, and
(iii) innovation that creates new products, devises new uses for existing products, and increases
the efficiency of input use. Analysis of sources of economic growth finds that the biggest
differences between developed and developing economies are in innovation performances.
Innovation is critical for economic growth, but it also becomes increasingly important
for addressing major development challenges, such as the ones related to inclusion and
sustainability.

Recognizing this pattern, many countries are attempting to promote innovation and
entrepreneurship. Market and government failures and other bottlenecks impede innovation
and entrepreneurship, particularly in developing countries. These countries need to build
the capacity to find, absorb, and use new technologies and processes as well as foster
entrepreneurs who can take risks, look for finance, and bring new products and processes
to market. The most important source of innovation in developing countries involves the
adaptation of technologies and processes that exist elsewhere but may be new to the country
or firm. However, innovation can also come from local efforts, with many low- and middle-
income countries becoming important sources of incremental innovation. Given its global
role, developmental mandate, and combination of public and private sector expertise, the
World Bank Group is uniquely positioned to play an important role in helping countries build
their innovation capabilities.

The Independent Evaluation Group (IEG) identified an investment portfolio of $18.7 billion in
innovation and entrepreneurship interventions over the past decade across the World Bank
Group. IEG found that this investment is substantial, but its effectiveness can be enhanced
through broad, systemic efforts on a set of complementary actions. At the corporate level,
the Bank Group has to articulate a clear vision of how innovation will be used to solve major
development problems and how this vision can be transformed into workable solutions. Given
the rapidly changing development context, urgent action is required to enhance coordination,

consultation, or linkages on innovation and entrepreneurship initiatives across networks, sectors, and regions, as well as across the Bank Group institutions.

Another challenge is to develop practical solutions for people who earn less than $2 a day. This is not a low-income country agenda but one that is also relevant for middle-income countries with large segments of their populations living in poverty. Sustained efforts are required to experiment with different mechanisms and implementation arrangements. Also important are monitoring and evaluation systems to facilitate scale-up of promising interventions and mechanisms to effectively capture and share knowledge from operations within and across the Bank.

Evaluation Context

An innovation is the implementation of a new or significantly improved product, a new process, a new marketing method, or a new organizational method in business practices, workplace organization, or external relations. Entrepreneurship refers to firms undertaking risks and marshaling resources in pursuit of perceived business opportunities. This evaluation focuses on innovative entrepreneurs, defined as new or existing innovative firms that organize a business and provide something new. Innovation and entrepreneurship can be mutually reinforcing and together can be a powerful source of improved productivity and competitiveness.

In emerging economies and developing countries there is a growing realization that innovation is a prerequisite for maintaining competitiveness in global markets and is critical for catching up with top performing emerging market economies and developed countries. The World Bank Group has supported lending and non-lending activities in science, technology, innovation, and enterprise development for more than three decades. Yet this evaluation is the first comprehensive review to span the broad range of strategies and policy instruments that Bank Group institutions have used to support innovation and entrepreneurship.

The conceptual framework that guides this evaluation is developed from theoretical literature on innovation and entrepreneurship processes, as well as from the Bank Group's own experience with respect to interventions that respond to market and government failures; it is also based on other bottlenecks that impede innovation and entrepreneurship. This conceptualization of innovation and entrepreneurship focuses attention on four areas of interventions: (i) support to research and development (R&D), (ii) strengthening entrepreneurial capabilities, (iii) providing financing schemes, and (iv) fostering linkages among the actors in innovation systems—all within the context of a broad enabling environment.

The overarching question that IEG seeks to answer through this evaluation is: To what extent did targeted Bank Group interventions foster innovation and entrepreneurship intended to transform new ideas into greater competitiveness, economic growth, and poverty reduction? Specifically, it assesses the quality of strategic guidance for selecting and designing interventions in support of innovation and entrepreneurship, examines the effectiveness of different types of interventions, and distills lessons to enhance the effectiveness of the Bank Group support for innovation and entrepreneurship. Evidence for the evaluation comes from a portfolio review of World Bank Group projects, staff interviews, field surveys, and desk reviews of documents from within and outside the Bank Group.

Emerging economies and developing countries have increased their investments in R&D capability to enhance their innovation potential. Although investments in R&D themselves do not automatically lead to innovation, the recent focus on building up R&D capability is indicative of the changing geography of innovation potential. Several studies corroborate the impact of rising R&D investments on productivity, exports, and to some extent job creation and poverty reduction.

Bank Group Rationale for Supporting Innovation and Entrepreneurship

The rationale for supporting innovation and entrepreneurship projects has evolved. In the 1990s, correction of market and government failures provided the major justification for innovation and entrepreneurship projects. This perspective led to a stream of projects that emphasized investments in public research infrastructure, improvement in the efficiency of R&D systems, and efforts to help the private sector commercialize products from R&D. More recent projects have a much broader perspective, with increasing focus on strengthening sector and firm-level competitiveness, diversifying from factor-driven to innovation-driven growth, and inclusive innovation.

Strategies to Support Innovation and Entrepreneurship

Innovation measures show that innovation outputs and inputs are strongly correlated with income levels. On average, innovation performance is stronger in high-income countries than in middle- and low-income countries. In many developing countries, innovations may come from knowledge and technologies from foreign sources or other users in the economy, or innovation may be created by domestic research from public institutions, universities, and private firms.

Strengthening innovation capacity has been an important factor in countries that have experienced rapid and sustained economic growth. Emerging economies and developing countries seeking to pursue development strategies that foster growth must build the capacity to acquire, disseminate, and use technologies to promote innovation and encourage new and existing firms to invest in business opportunities. The Bank Group can play a vital role in helping countries build their innovation capacity. No single path to innovation drives development; experience shows that countries have used a variety of strategies to foster innovation and entrepreneurship.

Within the diversity of paths to innovation, a review of the experience of countries at different stages of development identified five common principles that can be useful in promoting innovation and entrepreneurship: (i) support public investment in R&D that focuses on improving efficiency and relevance to end users as well as strengthening the use of research results in public policy decisions; (ii) build domestic science, technology, and innovation capabilities to make effective use of global knowledge; (iii) strengthen linkages between public R&D and private sector users of technology and knowledge; (iv) build a strong enabling environment, including an effective use of information and communications technology; and (v) provide flexible financing arrangements to encourage innovative firms to undertake risks to develop new products, processes, and services.

Although various Bank Group strategies have signaled that support to innovation and entrepreneurship is or should be a priority, none has articulated a vision for the Bank Group's engagement. Moreover, some sector strategies include innovation and entrepreneurship as a priority, but efforts to coordinate, consult, or link work across sectors or between institutions are rare. So far, the Bank Group has not articulated an integrated strategy to support innovation and entrepreneurship at the country level. Bank Group Country Assistance Strategies (CASs) reflect increasing demand for innovation projects across different income categories. Upper-middle-income countries, such as Brazil, Chile, and China, give high priority to innovation in their development plans. CASs for lower-middle-income and low-income countries also prioritize innovation and entrepreneurship to improve competitiveness, as well as diversification from resource-based to knowledge- or innovation-driven development.

Supporting Innovation and Entrepreneurship in Projects

The World Bank Group institutions have a significant and diversified portfolio of activities to support innovation and entrepreneurship. This includes lending and investment and advisory services initiatives such as the Development Marketplace at the World Bank Institute and infoDev.

World Bank lending for innovation and entrepreneurship is directed toward governments. IEG identified a lending portfolio of 119 innovation and entrepreneurship projects between FY00 and FY12. Total lending for these projects was $8.2 billion. Of these projects, 106 directly supported innovation and entrepreneurship activities; they totaled $4 billion. Support for such projects, once concentrated in middle-income countries, is increasingly found in lower-income countries.

The International Finance Corporation (IFC) supports innovation and entrepreneurship by companies, investing directly in start-ups and existing companies. IEG identified 300 innovation and entrepreneurship projects in client companies between FY00 and FY11. Total investment commitment in these projects was $5.7 billion, with $4.9 billion in loans and $0.8 billion in equity. Innovation projects were concentrated in lower-middle-income countries, about two-thirds of them in China and India.

The Multilateral Investment Guarantee Agency (MIGA) provides political risk guarantees to companies. Between FY00 and FY12, MIGA supported 108 innovation-related projects, issuing $4.8 billion in guarantees. These projects accounted for 24 percent of the total volume of guarantees issued during the period. About half of the volume of investment guarantees issued was in low-income countries.

Four types of targeted interventions have been used to support innovation and entrepreneurship within a broad-based enabling environment. These interventions to support innovation and entrepreneurship are:

• Support for public and private R&D

• Strengthening entrepreneurial capabilities

• Financial support for early-stage start-ups

• Fostering linkages between actors in the innovation system.

Interventions supporting R&D infrastructure—funding for R&D; intellectual property rights regime; national quality infrastructure; capacity building for science, technology, and engineering; and building entrepreneurial capabilities, particularly skills development and training in small and medium-size enterprises (SMEs)—were the top interventions the Bank supported. Enterprise-based support has helped establish new knowledge-based companies, including innovative SMEs and incubators.

Bank projects used various mechanisms to implement interventions that support innovation and entrepreneurship. Competitive research grants have been used to improve performance in public research systems, improve research-industry linkages, and promote private sector participation in public sector research. Matching grants have been used to facilitate development of new products through collaboration between firms and R&D institutions, thereby providing incentives for firms to bring innovations to market. This mechanism has also been used to help entrepreneurs finance the cost of business development services, export promotion activities, and technology upgrading. By providing a range of business support and services, incubators have been used to help firms grow and bring innovations to market.

Recently the World Bank and the Organisation for Economic Co-operation and Development developed the Innovation Policy Platform to foster the use of innovation policies and programs to increase sector- and firm-level competitiveness across industries and countries.

IFC's innovation and entrepreneurship projects focus almost exclusively at the firm level. Interventions aim to strengthen incentives for firm-level growth through technology transfer and diffusion, upgrading existing technologies and processes, and introducing new products and services. IFC also provides financial support for early-stage start-ups, mainly through investments in venture capital funds.

The majority of IFC interventions support technology upgrades mainly through helping firms procure new technology or processes or acquire new production technology and know-how through technology exchange or diffusion. IFC's financing for this is mainly in the manufacturing sector. Innovations also occur when new products, processes, or marketing or organizational models are introduced into the markets. These interventions are dominated by transactions in financial markets.

In cases where capital markets may not provide long-term capital for risky ventures with uncertain outcome, IFC has invested in venture capital funds that focus on early-stage companies and innovative SMEs. These investments have provided start-up firms with equity capital as well as managerial expertise, market information, and other forms of technical assistance.

The main channels through which MIGA's support for foreign direct investment fosters upgrades to technology in client firms were technology transfer and acquisition of new production technology and processes. The bulk of technology upgrade interventions were in infrastructure. MIGA also supported client firms in introducing new products and processes in the market, including support for the establishment of new financial institutions.

The World Bank Group has also supported analytical work, mainly through analytical and advisory activities at the Bank and through Advisory Services at IFC. At the Bank, analytic work has focused on innovation policy, knowledge economy, and technology, mainly to inform government policies. The Bank's technical assistance emphasized strengthening institutions and country capacity to implement innovation projects. IFC Advisory Services provided support to build entrepreneurial capabilities and provide management training and skill development as well as institution building or policy reform. Interventions focusing on building capabilities in start-ups and innovative SMEs were the most prevalent.

Performance of Bank Group Support

WORLD BANK PROJECT PERFORMANCE

IEG reviewed 64 closed World Bank projects that supported innovation and entrepreneurship. Most Bank projects had objectives to increase competitiveness through innovation or technology development. About 80 percent of completed projects had satisfactory project outcomes. Their performance was slightly higher than for other Bank projects evaluated during this period, but that difference was not statistically different. Innovation and entrepreneurship projects were less successful in low-income countries than in lower-middle- and upper-middle-income countries. Such differences suggest that the performance of interventions depends on the local context.

Of the 64 projects, 35 were considered major innovation and entrepreneurship projects because more than 50 percent of Bank costs were specifically allocated to innovation and entrepreneurship activities. Major innovation and entrepreneurship projects were significantly more successful in achieving their relevant objectives than minor innovation and entrepreneurship projects. Furthermore, the difference in successful achievement of project objectives between major and minor innovation and entrepreneurship projects holds across country income level.

The quality of Bank project design and supervision, and of borrower performance in preparing and implementing projects, is a key determinant of project outcomes. Bank and borrower performance in innovation and entrepreneurship projects, compared with other Bank projects, was significantly less successful in low-income countries and better in upper-middle-income countries. The main problems with project performance associated with the Bank's role, irrespective of whether projects achieved their objectives, were mostly related to design issues (complex design, unrealistic targets, inadequate monitoring and evaluation) and quality of supervision. On the borrower side, problems were caused by inadequate performance of government and implementing agencies, often resulting in implementation delays.

The quality of Bank supervision is important, but project supervision and design have complementary effects. A project that is not properly designed is less likely to achieve its objectives even with good supervision. However, a good design is not enough to ensure the achievement of project objectives when the project is poorly supervised. Appropriate targeting and effective monitoring and evaluation also work together.

Project components related to support to public R&D were mostly successful—77 percent satisfactory or better rating—in achieving component objectives. Project components fostering linkages between research and industry had the highest ratings, but there were relatively few of these activities. The performance of various innovation and entrepreneurship interventions reflects an emphasis on correcting market and government failure in innovation projects that were approved in the 1990s or earlier.

Competitive research grants have been used effectively to improve efficiency in public research systems and universities, strengthen research and industry collaboration, and encourage publicly funded standards boards to respond to industry demands. However, competitive grants have not been effective in dealing with reforms in public sector agencies or targeting poor farmers or regions without additional support to help build their capacity.

Matching grants helped improve the performance of entrepreneurs and provided incentive for firms to take innovations to market. The effectiveness of this mechanism was limited by design and implementation problems around uncertainties with eligibility criteria, slow and costly implementation, low uptake, complex processing, reimbursement issues, budgetary procedures, and political interference.

Incubators have been used to support entrepreneurs with a view to increase survival rates for start-ups and innovative SMEs. Available evidence suggests that the success of business incubators is mixed. Evidence on the effectiveness of World Bank Group–supported incubators is limited because there have been relatively few evaluations that assessed the performance of firms that exit the incubator compared to those that did not use its services. An analysis of lessons from infoDev's support for business incubators suggests positive effects on project outputs and outcomes. However, there is not much that can be said about the impact of infoDev's incubation interventions on the basis of this study because it did not specify useful comparisons and benchmarks.

IFC PROJECT PERFORMANCE

The performance of IFC investment projects was assessed on three measures: investment outcome, project business success, and private sector development impact. Assessment of project performance was based on 203 projects—out of the 300 projects reviewed—with evaluated evidence from expanded project supervision reports. Among projects evaluated during the same period, the cohort of interventions that supported firm-level innovation had significantly lower development and investment outcome success ratings than other projects. In addition to the inherently higher risks of innovative projects, the relatively lower performance ratings for innovation and entrepreneurship projects are associated with lower ratings in work quality, particularly screening, appraisal, and structuring. Given the higher-risk profile of innovation projects, their performance needs to be assessed also on a portfolio basis. On a portfolio basis, the average financial and economic rate of return on innovation-related projects was just as good as for projects without innovation components.

In IFC's innovation and entrepreneurship projects, issues in three areas accounted for the majority of problems associated with partly unsuccessful or lower outcomes sponsors, markets, and risk. Given the high risks associated with innovation-related projects, there is a likelihood that IFC may have identified these issues but underestimated their implications on development outcomes. Implementation setbacks were encountered in projects regardless of their development outcome ratings. Sound market analysis was critical in ensuring that IFC's innovation and entrepreneurship projects were effective in achieving their development outcomes.

Across innovation- and entrepreneurship-related projects, IFC's support for investment in technology upgrading through technology transfer, diffusion, or technology acquisition had the highest proportion of projects with successful or better ratings for development outcomes, returns to IFC, and private sector development. In contrast, financing schemes that supported early-stage start-ups, R&D for firm-level capacity, and establishment of financial institutions had a relatively low proportion of projects with successful or better ratings on these key performance indicators. These low performance results, though indicative of broader performance drivers and obstacles, should be interpreted with caution because they are based on relatively small samples.

MIGA PROJECT PERFORMANCE

MIGA's development performance was based on a small sample of evaluated projects, making it difficult to draw robust inferences. These projects were assessed based on development outcome, business performance, economic sustainability, and private sector development. Nine of the 18 evaluated innovation and entrepreneurship projects had development outcome ratings that were satisfactory or better. A similar number of projects had business performance ratings that were satisfactory or better, but a higher proportion had satisfactory or better private sector development effects. Twelve of the 18 evaluated projects had satisfactory or better economic sustainability ratings, indicating positive welfare effects of these projects on society and stakeholders.

Even though the MIGA sample is too small to draw stronger statistical inferences, an analysis of development outcome indicators by types of interventions provides some interesting insights. For example, the majority of projects that supported firm-level upgrading through the introduction of new products and processes into markets had development outcome ratings that were satisfactory or better. IEG found many cases where MIGA's support for firm-level technology upgrading through technology transfer, technology diffusion, and acquisition of new technology helped promote innovation, skill development, and growth of the private sector.

The quality of MIGA's assessment, underwriting, and monitoring had the lowest successful ratings. Assessment, underwriting, and monitoring are front-end work that can have the greatest influence on a project's success.

Learning from Bank Group Interventions

Going beyond project performance ratings to draw broader lessons from project evaluation across Bank Group interventions provides additional evidence that some mechanisms have been particularly effective in helping provide development solutions. World Bank agricultural research projects, for instance, have used competitive research grants effectively to increase the efficiency and relevance of public research institutions. IFC and MIGA technology transfer projects have been effective in increasing firm-level productivity and competitiveness by providing access to state-of-the-art technologies and knowledge flows.

But challenges remain, particularly in areas where the Bank Group does not have a long history of operations or experience using specific mechanisms, such as matching grants, business incubators, venture capital, and financing for the introduction of new products, processes, and institutions. The evidence suggests that the Bank Group has had limited success using these mechanisms to implement innovation and entrepreneurship interventions.

Bank Group staff identify lessons and develop and use tacit knowledge on innovation and entrepreneurship in the course of their work. Although most staff share project experience, the flow of knowledge on innovation and entrepreneurship is limited, particularly across networks, sectors, and regions. The limitations of knowledge flows are amplified across the Bank Group, given that staff are less inclined to share their experiences and lessons. Given that tacit knowledge is dominant on thematic issues like innovation and entrepreneurship, the limited flows of knowledge among staff within and across the three institutions imply considerable inefficiencies that limit the effectiveness of Bank Group support to operational teams and country clients.

Conclusions and Recommendations

World Bank Group interventions have helped developing countries build their innovation capacities, but the Bank Group's effectiveness could be enhanced by adopting a more strategic approach to supporting innovation and entrepreneurship for development.

The policy rationale for supporting innovation and entrepreneurship in Bank Group projects has moved away from a narrow focus on market and government failures toward a much broader perspective that integrates other bottlenecks impeding innovation and entrepreneurship. When looking at the achievement of stated objectives and outcomes or benchmarks, evaluative evidence suggests that World Bank-supported innovation and entrepreneurship projects perform as well as other Bank Group projects. On a portfolio basis, IFC's innovation-related projects performed just as well as projects without innovation components, generating financial and economic returns that were above IFC's benchmarks. The limited evaluated projects supporting innovation and entrepreneurship performed just as well as other MIGA projects.

Several financing and non-financing mechanisms have been used to implement these interventions. IEG found mixed evidence on the performance of some mechanisms. In general, interventions are more likely to perform well in areas where the Bank Group has operational experience.

However, current corporate and sector strategies do not provide adequate guidance on how to develop interventions that can help client countries select, design, and implement policies and integrated programs to support innovation and entrepreneurship in a holistic manner. In fact, the World Bank Group does not have a comprehensive strategy and results framework for projects supporting innovation and entrepreneurship.

This is partly because the agenda on innovation and entrepreneurship is still evolving. Bank Group interventions in this field have tended to be articulated around other thematic areas and not necessarily around innovation and entrepreneurship as a theme.

Analysis of Bank Group interventions from country perspectives shows that they are often designed and implemented at the sector level, with strong alignment to institutional experience and specialization. These efforts do not address the systemic nature of innovation that is required for solving development challenges at the country level. In addition, limited mechanisms and weak incentives to share learning from design and implementation have restricted knowledge flows among sectors, regions, and Bank Group institutions. Yet there is increasing client demand for work on innovation and entrepreneurship. This needs to be better reflected and integrated across Bank Group operations and analytical work. Such efforts will help in improving the effectiveness of work on the ground and articulating a consistent set of messages to clients.

IEG presents the following recommendations to strengthen the effectiveness of Bank Group support for innovation and entrepreneurship.

There is a myriad of activities on innovation and entrepreneurship within the Bank Group but few formal efforts to coordinate, consult, or link these activities. A well-coordinated cross-sectoral set of actions needs to emerge from different Bank Group activities on innovation and entrepreneurship. Going forward, there is need for better planning, joint decision making, improved coordination, and quality control of the Bank Group's work in this arena.

RECOMMENDATION 1: The Bank Group should develop and implement a consistent and well-coordinated strategic framework that highlights the relationships between work on innovation and entrepreneurship across different sectors and institutions. This framework should consider the context of the new Bank Group strategy and provide the building blocks for developing innovation strategies, policies, and programs that will help client countries strengthen innovation-driven growth.

• The Finance and Private Sector Development (FPD) Network has an explicit practice that focuses on innovation and entrepreneurship, so it is well placed to provide the multisectoral coordination that such an effort demands.

The World Bank and IFC have provided financial support for early-stage start-ups through venture capital funds as well as loans and grants to innovative and entrepreneurial companies and SMEs. World Bank financing support for start-ups has mainly focused on matching grants and a few projects have included venture capital funds. Relative to the Bank, IFC has invested more in venture capital funds and other private equity funds that focused on early-stage and innovative firms. World Bank and IFC financing for early-stage start-ups has had mixed results, and there is need for a more systematic assessment of performance drivers and obstacles. There is an urgent need to understand the conditions under which venture capital funds and other types of risk financing are likely to be successful, particularly in developing countries that have limited funding opportunities for early-stage financing. In such contexts effective support for start-ups should consider issues such as investment capital (seed capital, minigrants) at early stages of enterprise formation, weak or non-existent markets, limited deal flows, policy dialogue, and financing regulations.

RECOMMENDATION 2: The World Bank and IFC should assess, develop, test, and learn from alternative approaches to provide risk financing for early-stage start-up firms that are at different stages of commercial growth.

• This is a fruitful area for collaboration between the World Bank and IFC, building on their respective comparative advantage.

- The discussion of financing for early-stage start-ups should not be done in isolation but embedded within an overall discussion of Bank Group support for innovation systems. Risk financing for early-stage start-ups should consider systemic and long-term conditions that are required for financing entrepreneurs in different stages of maturity within innovation systems.

The Bank Group, particularly IFC and MIGA, supports technological upgrading activities in firms through technology transfer, diffusion, upgrading of technologies and processes, and introduction of new products, processes, and business models. These interventions have provided important sources of innovation in firms and countries. Such efforts need to be strengthened and made more systematic to enhance learning and knowledge flows between and across firms and countries.

RECOMMENDATION 3: The World Bank, IFC, and MIGA should take proactive steps to distill, document, and facilitate knowledge sharing on approaches to facilitating innovation from technology transfer, diffusion, and upgrading of technologies.

Much of the Bank's work on innovation and entrepreneurship is concentrated in lower- and upper-middle-income countries. But innovation is important at all stages of development and clients from low-income countries are increasingly requesting Bank support for projects that address challenges specific to developing country contexts. Countries such as China and India have become significant actors in inclusive and incremental innovation that can be scaled up to other developing countries. Thus, there are promising opportunities to foster inclusive innovation through South-South interactions. The Bank Group needs to make a special effort to develop innovation and entrepreneurship projects that address various aspects of innovation that benefit poor and other underserved populations in low- and middle-income countries.

RECOMMENDATION 4: The World Bank Group should broaden its involvement in inclusive innovation projects in response to client demands. The World Bank and IFC should intensify current efforts to pilot, assess, learn, and scale up inclusive innovation projects with partners.

- The Bank should focus on building innovation capacity early in the development process to help low-income countries acquire and adapt the types of innovation that address challenges that are specific to their local contexts.

- Teams developing inclusive innovation projects should pilot, assess, and scale up different types of inclusive innovation. Such efforts must be underpinned by an effective monitoring and evaluation system so that the learning process can inform dissemination and use of new products, processes, and services in other development contexts.

The World Bank Group's support to innovation and entrepreneurship has not been tracked very well. It has a diversified portfolio of activities that can provide good learning opportunities to foster innovation and entrepreneurship. However, much of the Bank's learning and knowledge is embodied as tacit knowledge that is often transferred through direct individual interaction. There is limited flow of knowledge on innovation and entrepreneurship across sectors, networks, and regions, as well as across Bank Group institutions. This leads to reliance on learning by doing, which is costly and limits effective utilization of Bank Group learning to devise efficient innovation policies and programs. The joint World Bank–Organisation for Economic Co-operation and Development Innovation Policy Platform provides a mechanism that can facilitate knowledge exchange, including tacit knowledge.

RECOMMENDATION 5: Consistent with ongoing World Bank Group knowledge reform, the FPD network at the World Bank needs to develop cost-effective and easily accessible procedures for codifying and disseminating information on project design and implementation experiences from its work on innovation and entrepreneurship. Similar efforts should be developed and implemented by IFC and at MIGA.

Project performance ratings suggest that innovation and entrepreneurship projects have mostly been successful. But there is mixed evidence on the effectiveness of key intervention and mechanisms that have been used to support innovation and entrepreneurship. A major problem in most of the monitoring and evaluation (M&E) information reported in project documents is that the most meaningful aspects of innovation and entrepreneurship are not measured. The few indicators reported focus mainly on R&D inputs but these do not capture innovation—new products, processes, and business models that are brought to the market. M&E of innovation policies and programs is critical to identify what kinds of policies and mechanisms are effective in specific contexts as well as improving the efficiency of resources allocated to innovation and entrepreneurship.

RECOMMENDATION 6: The World Bank and IFC should identify innovation projects involving incubators, matching grants, venture capital, and other risk financing interventions that can be assessed to facilitate learning and scaling up of those that are promising.

• Teams working on innovative projects at World Bank and IFC should build robust M&E into the design and implementation of these interventions.

Management Response |

World Bank Group management welcomes this Independent Evaluation Group (IEG) review of innovation and entrepreneurship. With the interest on innovation policy increasing and the demand for assistance from client countries growing exponentially in recent years, this IEG report could not come at a better time. The global financial crisis required developing economies to actively seek new sources of economic growth. Natural resource-intensive economies are pursuing new ways to diversify their productive capacity to build resilience to commodity cycles. Middle-income countries (MICs) are looking for ways to escape the "middle-income trap." A number of developmental challenges, from climate adaptation to food security and health, require new, efficient technological solutions adapted to developing countries' needs. In such context, innovation and entrepreneurship are increasingly seen as essential ingredients for economic and social prosperity.

World Bank Management Comments

World Bank management welcomes this evaluation. The report is particularly important for the Innovation, Technology, and Entrepreneurship (ITE) Global Practice. The two-year-old ITE Practice was created by the Financial and Private Sector Development (FPD) Network as part of a new business model pilot in the World Bank and a response to the rising demand from client countries. It seeks to better integrate the various products and instruments available within the World Bank Group to provide a more programmatic and comprehensive approach to client service, geared toward achieving tangible development impact in the form of new investments, jobs and income opportunities for the poor. The Practice integrates staff from FPD, as well as from other World Bank networks and the International Finance Corporation (IFC) collaborating on the innovation and entrepreneurship agenda. The Practice welcomes the evaluation as a unique opportunity to learn from a systematic and careful review of World Bank Group initiatives within the Practice domain.

BROAD CONCURRENCE WITH ANALYSIS AND CONCLUSIONS

Management concurs broadly with the findings and conclusions of the evaluation, in particular (i) the endorsement of innovation and entrepreneurship as a prerequisite for maintaining competitiveness in global markets and a critical factor for growth of developing economies; (ii) the finding that articulating World Bank interventions supporting innovation and entrepreneurship around various thematic areas (such as agriculture and rural development, private sector development, education, and information and communications technology) in the absence of a strategic framework for innovation and entrepreneurship may affect their impact; (iii) the conclusion that projects primarily focused on innovation are more likely to succeed than those just incorporating some components—and useful evidence of the importance of conveying some level of comprehensiveness in project design, discouraging piecemeal interventions; (iv) the notion that innovation policies must be placed in the context of other productivity enhancing policies and should not be considered in isolation from market conditions; and (v) the recognition that no single path to innovation drives development and that countries have used a variety of strategies to foster innovation and entrepreneurship.

OTHER OBSERVATIONS

The report does not take a comprehensive look at World Bank Group initiatives supporting entrepreneurship. It makes a deliberate choice to limit its analysis to innovative entrepreneurship with an emphasis on start-ups. A comprehensive framework on entrepreneurship should also consider entry, growth, and exit factors and should not only consider the strengthening of entrepreneurial capabilities, but also incentives.

Among World Bank interventions, the report overemphasizes the role of public investment in research and development relative to other types of institutions, incentives, and initiatives that have contributed to the dissemination of technologies and their adoption by firms and producers.

The report indicates that the Bank Group does not have a long history of operations using certain mechanisms, among other things, incubators and financing, to support the introduction of new products and processes and has had limited success using these. Although we agree that the monitoring and evaluation framework for these instruments needs to be strengthened (a conclusion of the evaluation), the report does not provide concrete evidence indicating that these interventions were not successful.

The report notes there are limited mechanisms in place to share learning about innovation and entrepreneurship and few formal efforts to improve coordination. However, important initiatives are already under way to foster learning, cross-fertilization, and codification of knowledge to support innovation policy. The ITE Global Practice, for example, brings together innovation and entrepreneurship specialists across the Bank Group in knowledge exchange forums and other learning activities. The Innovation Policy Platform (IPP), under advanced development, will provide a repository of knowledge on the "how to" of innovation policy and a collaborative space for users (among other things, World Bank Group staff, policy makers, practitioners, and analysts) to exchange knowledge. The IPP (a joint initiative with the Organisation for Economic Co-operation and Development) is being led by FPD with initial support from the World Bank Institute and will involve developers and users across the Bank Group. The module on innovation for agriculture, for example, was developed with support from the agricultural and rural development group.

World Bank management appreciates the report's recommendations, especially the need for a "Strategic Framework" that will highlight the relationship between work on innovation and entrepreneurship across sectors and different institutions. This framework will be developed taking into account the directions and institutional arrangements emerging from the ongoing Bank Group strategy formulation and the change process to ensure consistency with overall thrust of reforms, particularly in the area of global knowledge sharing.

We would also like to note several initiatives under way that are already helping address many of the recommendations, such as the aforementioned IPP; the development of new external partnerships, among other things, with the Global Research Alliance to assist client countries in the development and implementation of their inclusive innovation strategies; and the development of the Early-Stage Financing Facility under infoDev, which will help test new financing instruments to support start-ups.

IFC Management Comments

IFC welcomes IEG's evaluation of the World Bank Group support for innovation and entrepreneurship. The report comes at an opportune time, given that support for innovation is an important part of our engagement strategy, particularly in MICs.

IFC supports innovation through its investment and advisory services in firms that develop, introduce, or adapt new products, approaches, or business models to increase efficiency and competitiveness, and therefore drive growth. These interventions spread across the value chain encompassing manufacturing, agribusiness, services, and other sectors beyond information, communication, and technology. IFC also provides advisory services aimed at creating an enabling environment conducive to the development of innovation through institution building and policy reform. In entrepreneurship, IFC provides advisory services on management education and skills training for micro, small, and medium-size enterprises to build entrepreneurial capabilities and investment services in venture capital funds to support early-stage start-ups.

Management appreciates the report's valuable independent evaluation of a cohort of IFC innovation and entrepreneurship interventions identified through a manual validation of projects that featured relevant key words. The report covers 300 investment projects committed between FY00–11, 203 of which have Expanded Project Supervision Reports, and 84 advisory projects undertaken between FY05 and FY12. To sharpen the focus on entrepreneurship and keep its relevance to innovation, the report rightly limits its coverage of entrepreneurship to "innovative entrepreneurs," which it defines "as new or existing innovative firms that organize a business and provide something new." It also excludes investment climate advisory and some entrepreneurship projects of the Bank and IFC that focus on small and medium-size enterprise capacity building. Given these scope limitations and the challenges in manually identifying projects, the report generated a reasonable cohort of relevant investment and advisory service projects in innovation and innovative entrepreneurship with an emphasis on start-ups.

The report correctly acknowledges that projects that include innovations are inherently riskier. Despite this, the development and financial performance of projects with innovation components held up well on a portfolio basis relative to those with no innovation components. The different level of performance on a project count basis reflects the risk/reward profile of innovative projects, especially innovative entrepreneurship. This is evident in the ten venture capital funds that support early stage start-ups, which display a different performance track than the rest of the evaluated IFC investments. The report recognizes that venture capital investments typically have high rates of failure in which only one or two investments earn high returns for every ten investments made. It states further that market practice assesses a fund's performance on a portfolio basis against its peers of the same vintage, not on a stand-alone basis. The positive effects on private sector development of IFC investments in venture capital funds confirm that this type of financing can be an important mechanism for fostering innovation, entrepreneurship, and growth of private enterprises. IFC will continue to operationalize lessons of experience in innovation and entrepreneurship to further enhance overall outcomes.

We agree with all six interrelated recommendations. The recommendation to develop a World Bank Group strategic framework in innovation and entrepreneurship provides an overarching platform for operationalizing the other five recommendations. Management will take a comprehensive view in developing specific approaches to operationalizing all of the report's six recommendations as part of formulating a Bank Group strategic framework for innovation and entrepreneurship.

MIGA Management Comments

The Multilateral Investment Guarantee Agency (MIGA) welcomes the IEG evaluation report of the World Bank Group support for innovation and entrepreneurship activities and finds it important and useful. It comes at a time of increased emphasis on synergies across the World Bank Group under the broad rubric of one World Bank Group. Given MIGA's mandate to facilitate the flow of productive cross-border investments into developing countries, the projects supported through its guarantee operations play a key role in the acquisition and transfer of technology as well as knowledge that is at the heart of innovation and entrepreneurship. This IEG evaluation has done a good job of capturing the contributions of MIGA-supported projects that promote innovation and entrepreneurship.

The evaluation findings are based on 18 MIGA projects that supported innovation and entrepreneurship, and as correctly noted in the IEG report, do not provide a robust sample for statistical analysis. However, management appreciates the effort made in the report to draw meaningful insights from these projects. In this regard, management notes that it would have been useful to include the evidence presented in IEG's (2012) Country Program Evaluation on Afghanistan, which includes a MIGA project (one of the 18) and illustrates well the complementarities of the World Bank Group in the information and communication technology sector (http://ieg.worldbankgroup.org/content/dam/ieg/afghanistan/afghan_eval _full.pdf; for example, Box 5.2: Synergies Among World Bank Group Institutions).

The report correctly acknowledges the role played by MIGA's political risk insurance product in addressing incentive problems that may cause firms to under-invest in innovative products and processes. Management also finds useful the report's identification of the main channels for technology upgrading in MIGA-supported projects as (i) technology transfer and (ii) acquisition of new production technology and processes. The report goes on to identify these two channels, together with capacity building (through training or knowledge transfer), as the mechanism by which MIGA's support for innovation and entrepreneurship leads to firm growth and expansion. Management notes these findings as useful from an operational standpoint, especially for purposes of articulating the development impact of MIGA operations featuring complex technologies.

The report states that MIGA support helped jump-start private sector foreign direct investment in postconflict situations in Mozambique and Nicaragua. Management regards this important finding as a validation of MIGA's ongoing support to fragile and conflict-affected situations as a strategic priority as well as the increasing emphasis, illustrated well by the recent launch of the Conflict-Affected and Fragile Economies Facility in June 2013.

Management notes the statement in the report regarding the quality of MIGA's front-end assessment and underwriting work as being vital, given its potential influence on project success. However, management finds that the report presents no supporting evidence. The report then concludes that the effectiveness of MIGA's interventions to support innovation and entrepreneurship will be enhanced with improvements in the quality of its front-end work in assessment, underwriting, and monitoring. Management notes that despite lack of evaluative evidence to support this conclusion, MIGA takes the quality of its front-end work seriously and will continue to learn from experience for improving the effectiveness of its interventions.

With regard to recommendations, management agrees with the overall recommendation to develop and implement a consistent and well-coordinated strategic framework across different sectors and institutions.

Management Action Record

A Strategic Framework for Innovation and Entrepreneurship

World Bank Group activities supporting innovation and entrepreneurship are articulated in different strategies, lending and investment operations, and analytical work. However, there are few formal efforts to coordinate, consult, or link these activities across sectors, networks, and institutions. Current corporate and sector strategies do not provide adequate guidance on how to develop effective innovation interventions. Bank Group interventions in this field have tended to be articulated around thematic areas of interventions and not necessarily around innovation and entrepreneurship.

IEG RECOMMENDATION

The Bank Group should develop and implement a consistent and well-coordinated strategic framework that highlights the relationships between work on innovation and entrepreneurship across different sectors and institutions. This framework should be developed, considering the context of the new Bank Group strategy and providing the building blocks for developing innovation strategies, policies, and programs that will help client countries strengthen innovation-driven growth.

The FPD Network has an explicit practice that focuses on innovation and entrepreneurship, so it is well placed to provide the multisectoral coordination that such an effort demands.

ACCEPTANCE BY MANAGEMENT

World Bank Group: Agree

MANAGEMENT RESPONSE

World Bank Group management will develop a "Strategic Framework" that will highlight the relationship between work on innovation and entrepreneurship across sectors and different institutions. This framework will also guide the operationalization of the other five recommendations. The framework will be developed taking into account the directions and institutional arrangements emerging from the ongoing World Bank Group strategy formulation and change process to ensure consistency with the overall reform thrust, in particular when it comes to global knowledge sharing.

An increased and coherent support for innovation and entrepreneurship can help accelerate the achievement of the World Bank Group's twin goals of ending extreme poverty and promoting shared prosperity. The IEG report correctly acknowledges that innovation and entrepreneurship can be a powerful source of improved competitiveness, creating jobs and helping drive growth. The World Bank Group's approach to innovation and entrepreneurship has rightly evolved over time in response to client demands, opportunities, threats, and taking into account the comparative advantages of each institution separately and together as appropriate, across sectors, themes, and regions. Given the recent launch of the Bank Group's twin goals, the knowledge gained over the years in this space, and IEG's independent assessment of the Bank Group experience, it is an opportune time to develop strategic directions or a strategic framework for the World Bank Group's work on innovation and entrepreneurship.

The World Bank Group strategic framework will assess current client demands on innovation and entrepreneurship; identify priority products and opportunities and tools for further experimentation and learning with Bank Group products to better meet client demands; identify mechanisms for further strengthening collaboration across departments and institutions within the Bank Group; identify mechanisms for better disseminating knowledge within the Bank Group as well as to clients; and explore opportunities for leveraging World Bank Group capacities through external partnerships.

The strategic framework or strategic directions will also help guide the operationalization of recommendations 2–5.

The strategic framework will be informed by the upcoming World Bank Group strategy and the FPD strategy and other existing strategies such as the IFC's MIC Strategy, IFC Road Map Papers, infoDev's medium-term work program, and the Bank Group's strategy for information and communication technology.

Risk Financing for Early-Stage Start-Ups

IEG FINDINGS AND CONCLUSIONS

An incentive problem relating to innovation investments is that they are often risky, with uncertain outcomes. This inherent uncertainty of success results in limited financing. The Bank Group has provided financial support for early stage start-ups through venture capital funds as well as loans and grants to innovative and entrepreneurial companies and SMEs. World Bank financing support for start-ups has mainly focused on matching grants and a few projects have included venture capital funds. Relative to the Bank, IFC has invested more in venture capital funds and other private equity funds that focused on early-stage companies and innovative SMEs, providing them with equity capital as well as managerial expertise, market information, and other forms of technical assistance. World Bank Group financing for early stage start-ups has had mixed results and there is need for a more systematic assessment of performance drivers and obstacles.

IEG RECOMMENDATION

The World Bank and IFC should assess, develop, test, and learn from alternative approaches to provide risk financing for early-stage start-up firms that are at different stages of commercial growth.

This is a fruitful area for collaboration between the World Bank and IFC, building on their respective comparative advantage.

The discussion of financing for early-stage start-ups should not be done in isolation but embedded within an overall discussion of Bank Group support for innovation systems. Risk financing for early-stage start-ups should consider systemic and long-term conditions that are required for financing entrepreneurs in different stages of maturity within innovation systems.

ACCEPTANCE BY MANAGEMENT

WB: Agree

MANAGEMENT RESPONSE

Further experimentation with early-stage financing instruments is necessary, and this will be further addressed in the context of the Strategic Framework. The forthcoming Early Stage Financing Facility, under infoDev's management, will also help explore new financing mechanisms to support start-ups.

This, coupled with a strengthened monitoring and evaluation (M&E) framework, as proposed under recommendation 6, will allow learning from the impact of such risk financing instruments.

Moreover, the innovation financing module of the IPP, discussed further under management response to recommendation 3, will serve to codify and disseminate experiences with early-stage financing, whether funded by World Bank Group projects or other sources.

The recent appointment of a director in IFC for telecom, media, and technology and early-stage investments will support strengthened capabilities in this area for the IFC.

Information Sharing

IEG FINDINGS AND CONCLUSIONS

The Bank Group supported technological upgrading activities in firms and farmers through technology transfer, diffusion, upgrading of technologies and processes, and introduction of new products, processes, and business models. These interventions have provided important sources of innovation in firms and countries. However, evidence indicates that not much is known about how technologies are absorbed, assimilated, and utilized to maximize learning, innovation, and spillovers into the broader economy.

IEG RECOMMENDATION

The World Bank, IFC, and MIGA should take proactive steps to distill, document, and facilitate knowledge sharing on approaches to facilitating innovation from technology transfer, diffusion, and upgrading of technologies.

ACCEPTANCE BY MANAGEMENT

WB: Agree

MANAGEMENT RESPONSE

The IPP under advanced development will provide a depository of knowledge on the "how to" of innovation policy and a collaborative space for users (among other things, Bank Group staff, policy makers, practitioners, and analysts) to exchange knowledge. The IPP (a joint initiative with the Organisation for Economic Co-operation and Development) is being led by FPD with initial support from the World Bank Institute and will involve developers and users across the whole World Bank Group.

The Strategic Framework will explore Bank Group complementary opportunities for documenting and facilitating knowledge exchange on technology diffusion, upgrading, and transfer.

MIGA can provide periodic written pieces that look at best practices and aggregated lessons of experience in the area of technology transfer relating to MIGA-guaranteed projects.

Inclusive Innovation Projects with Partners

IEG FINDINGS AND CONCLUSIONS

Inclusive innovation is a relatively new concept but is getting increased attention by some developing country governments and development institutions including the World Bank Group. In more recent Country Assistance Strategies, upper-middle-income China and Brazil have focused on emphasized inclusiveness and requested Bank support for promoting inclusive innovation that addresses the needs of the poor.

IEG RECOMMENDATION

The World Bank Group should broaden its involvement in inclusive innovation projects in response to client demands. The World Bank and IFC should intensify current efforts to pilot, assess, learn, and scale up inclusive innovation projects with partners.

The Bank should focus on building innovation capacity early in the development process to help low-income countries acquire and adapt the types of innovation that address challenges that are specific to their local contexts.

Teams developing inclusive innovation projects should pilot, assess, and scale up different types of inclusive innovation. Such efforts must be underpinned by an effective M&E system so that the learning process can inform dissemination and use of new products, processes, and services in other development contexts.

ACCEPTANCE BY MANAGEMENT

WB: Agree

MANAGEMENT RESPONSE

In the context of the Strategic Framework, Bank Group management will explore options and opportunities for piloting more and scaling up initiatives on inclusive innovation as well as facilitating South-South knowledge exchange around successful inclusive innovation projects.

As part of this scaling-up and knowledge exchange effort, FPD East Asia has already developed a partnership agreement with the Global Research Alliance, which seeks to apply innovative science and technology to improve the lives and economic opportunities of the poor.

Codification of Information on Innovation and Entrepreneurship

IEG FINDINGS AND CONCLUSIONS

World Bank sector and theme codes do not use innovation, entrepreneurship, or related terms to report on Bank activities. Nor do IFC and MIGA have a system that officially records or tracks innovation. The World Bank Group's support to innovation and entrepreneurship has not been tracked very well and experience has not been shared systematically.

An IEG survey indicates that Bank Group staff develop and use tacit knowledge on innovation and entrepreneurship in the course of their work, but this knowledge does not adequately flow within and across the Bank Group, resulting in organizational inefficiencies and limiting the effectiveness of Bank Group support to clients. Team leaders indicated that it is quite challenging to capture best practices because there are no mechanisms or time allocated to extract and transmit lessons from operations over time.

IEG RECOMMENDATION

Consistent with ongoing World Bank Group knowledge reform, the FPD Network at the World Bank needs to develop cost-effective and easily accessible procedures for codifying and disseminating information on project design and implementation experiences from its work on innovation and entrepreneurship. Similar efforts should be developed and implemented by IFC and at MIGA.

ACCEPTANCE BY MANAGEMENT

WB: Agree

MANAGEMENT RESPONSE

Consistent with the overall approach to knowledge management emerging from the World Bank Group change process, the Strategic Framework will explore ways on how best to identify and disseminate lessons of experience within each institution and across the three institutions of the Bank Group.

In IFC, this will be informed by its approach in global knowledge sharing.

Since the establishment of the ITE Global Practice, an effort has been launched to track more systematically World Bank projects (lending, technical assistance, and Reimbursable Advisory Services) supporting innovation and entrepreneurship.

In addition, as noted under recommendation 3, the IPP will provide a depository of knowledge on the "how to" of innovation policy and a collaborative space for users (among other things, World Bank Group staff, policy makers, practitioners, and analysts) to exchange knowledge.

Monitoring and Evaluation of Innovation and Entrepreneurial Projects

IEG FINDINGS AND CONCLUSIONS

Very little is known about the impact of World Bank Group investments supporting innovation and entrepreneurship. Project performance ratings suggest that innovation and entrepreneurship projects have mostly been successful. However, most of the M&E information reported in project documents indicates that they do not measure the most meaningful aspects of innovation and entrepreneurship. The few indicators reported focus mainly on research and development inputs but these do not capture innovation-new products, processes, and business models that are brought to the market.

IEG RECOMMENDATION

The World Bank and IFC should identify innovation projects involving incubators, matching grants, venture capital, and other risk financing interventions that can be assessed to facilitate learning and scaling up of those that are promising.

Teams working on innovative projects at World Bank and IFC should build robust M&E into the design and implementation of these interventions.

ACCEPTANCE BY MANAGEMENT

WB: Agree

MANAGEMENT RESPONSE

The Strategic Framework will seek to provide guidance on how monitoring and evaluation of innovation and entrepreneurship initiatives can be strengthened and share lessons from ongoing initiatives (for example, technology extension, skills upgrading, and financing of entrepreneurship) that have already put in place rigorous M&E frameworks.

The monitoring and evaluation framework will also recognize that, in some cases, the full impact will only be observed several years after the completion of the intervention and closing of the project due to lagged effects.

Chairperson's Summary:
Committee on Development Effectiveness

The Sub-Committee of the Committee on Development Effectiveness considered an Independent Evaluation Group (IEG) evaluation entitled *World Bank Group Support for Innovation and Entrepreneurship* and draft management response.

Summary

The Committee welcomed the timeliness and broadly endorsed the findings of this first comprehensive review of the policies and instruments that the World Bank Group has used to promote innovation and entrepreneurship. They praised IEG and management for their constructive collaboration, with each maintaining its independence.

Members stressed innovation's critical linkages to economic growth and to the World Bank Group's twin goals of reducing extreme poverty and promoting sustained prosperity. They also appreciated that the Bank Group has an existing diverse and significant portfolio of lending and non-lending activities that support innovation and entrepreneurship. Members observed, in addition, that IEG's country-level analysis pointed to the need for a strategic "One World Bank Group" approach on innovation and entrepreneurship rather than isolated efforts in order to address the systemic nature of innovation and entrepreneurship. Members also agreed that the evaluation could provide input and direction on strategic priorities for the World Bank Group's new innovation technology and entrepreneurship global practice of the Financial and Private Sector Development Network, and, in turn, the World Bank Group strategy as a whole.

The Committee observed the evaluation's findings that World Bank Group–supported interventions on innovation performed equally well as those without innovation components, and that innovation and entrepreneurship go beyond the middle-income country agenda. They agreed that effective innovation policy is formed within the overall policy environment and urged management to share its timeframe for the innovation strategy framework and Innovation Policy Platform, which will provide knowledge on the "how-to" of innovation policy, and how to achieve cost-efficiencies in operations and will cover, among other things, such topics as financing of innovation, skills development, innovation for green growth, and inclusive innovation.

Members stressed the importance of scaling up inclusive innovation that focuses on goods, services, and delivery methods for those most in need. In response to the reported lower success rate of innovation and entrepreneurship projects in low-income countries, members urged management to include capacity building in the innovation and entrepreneurship strategy to create an enabling environment for entrepreneurship. They also supported the usefulness of greater institutional knowledge sharing on innovation and entrepreneurship approaches, which was a limitation highlighted by the evaluation. The Committee underscored the need for more effective monitoring and asked that greater efforts be allocated to dissemination of the lessons of innovative projects. Members stressed the need for innovative approaches to access to finance. One member raised the need for early-stage financing, to create an entrepreneur-friendly environment. Members reiterated the need for synergies within the World Bank Group in the risk-taking framework, particularly in the high-risk innovation and entrepreneurship environment.

Juan José Bravo
CHAIRPERSON

1 Evaluation Context

CHAPTER HIGHLIGHTS

- Innovation and entrepreneurship can be a powerful source of improved productivity and competitiveness, helping to reduce poverty and stimulate long-term economic growth.

- Innovation is the development or adaptation of new products, processes, or services. Innovative activities are not limited to research and development (R&D), but may include other forms of non-R&D interventions

- The focus is on innovative entrepreneurs, defined as new or existing innovative firms that organize a business and provide something new.

- This evaluation is the first comprehensive review of the World Bank Group's support for innovation and entrepreneurship that spans the range of policies and instruments that the Bank Group has used in its client countries.

- The overarching question of the evaluation is to what extent World Bank Group–targeted interventions foster innovation and entrepreneurship intended to transform new ideas into greater competitiveness, economic growth, and poverty reduction.

Innovation, the concrete application of knowledge for economic growth and improvement in social welfare, can be an important catalyst for solutions to systemic development challenges. It is defined as the implementation of a new or significantly improved product (good or service) or process, a new marketing method, or a new organizational method in business practices, workplace organization, or external relations (OECD and Eurostat 2005). Distinct from invention, which is the first conception of something new, innovation is what allows countries, firms, and individuals to get new, more, and better goods and services with less. A broad concept of entrepreneurship refers to firms undertaking risks and marshaling resources in pursuit of perceived business opportunities (Pasquier and Stone 2008; OECD 2008). Entrepreneurs serve as agents of change and growth in market-based economies, providing a major channel through which innovative ideas can be turned into wealth.

Innovation and entrepreneurship are important for economic growth (Srinivasan 2004). They can be mutually reinforcing and together can be a powerful source of improved productivity and competitiveness, helping to reduce poverty and stimulate long-term economic growth (Dutz 2007; World Bank 2010a).

In this evaluation, innovation is the development or adaptation of new products, processes, or services. It includes the transfer or adaptation of innovations from other societies or countries to local markets and the diffusion and adoption of innovations in an economy. At the firm level, innovation can occur in products, processes, marketing, and organizational arrangements, encompassing changes in firm's activities.[1] Thus, innovation activities are not limited to research and development (R&D), but may include other forms of non-R&D interventions such as technology diffusion, building innovation capabilities, and acquisition of supportive infrastructure such as information communications and technology (ICT).

Entrepreneurs are those persons and firms that seek to generate value through the creation or expansion of economic activity by identifying and exploiting new products, processes, or markets (OECD 2008). This evaluation focuses on innovative entrepreneurs, defined as new or existing innovative firms that organize a business and provide something new—a product, process, type of business structure, or approach to marketing.[2] The treatment of innovative entrepreneurship in this evaluation is narrower than a comprehensive approach that considers firm entry, growth, and exit.[3] It does, however, allow the Independent Evaluation Group (IEG) to focus attention on policy actions that can be used to foster innovation in new firms entering the market or existing innovative firms. Thus, throughout this evaluation the term entrepreneurship refers to new firms and existing innovative firms.

As long ago as 1934, Joseph Shumpeter argued that innovation is a necessity for economic growth and development, a point made more recently by Baumol (2002) and Aghion (2006).

Estimates of the contribution of innovation to economic growth and improvement to welfare vary, but it is widely acknowledged that it explains more than half of productivity growth and income variation across countries (Solow 1957).

In emerging economies and developing countries there is a growing realization that innovation is a prerequisite for maintaining competitiveness in global markets and is critical for catching up with top performing emerging market economies and developed countries (OECD and World Bank 2009).[4] Innovation is critical for firms to improve their competitiveness or deliver goods or services more efficiently to customers (OECD 2011).

The global financial crisis and subsequent economic downturn required developing countries to actively seek new sources of economic growth. Natural resource-intensive economies are pursuing new ways to diversify their productive capacities to enhance their resilience to commodity cycles. Emerging economies are also exploring new sources of growth to escape the "middle-income trap." Development challenges such as food security, climate change, disease, and energy use require new and efficient technological and organizational innovations that can be adapted to the needs of developing countries (World Bank 2010b). In such contexts, innovation and entrepreneurship are increasingly seen as essential ingredients for economic and social prosperity. Moreover, innovation that focuses on goods, services, and delivery methods relevant to the needs of people at lower income levels can support inclusive growth, providing consumption and employment opportunities for large segments of the population (Dutz 2011). Such initiatives, referred to as "inclusive innovation," are getting increased attention from some developing country governments and development institutions, including the World Bank Group.[5]

This realization is making countries as diverse as Armenia, Brazil, Mexico, and Mozambique turn to international organizations and think tanks such as the World Bank Group, the Organization for Economic Co-operation and Development (OECD), and the United Nations Conference on Trade and Development (UNCTAD). They seek investment and policy advice to support development strategies that will promote innovation and entrepreneurship in order to shift competitiveness and diversification from resource-based to innovation-driven strategies (OECD and World Bank 2009; UNCTAD 2011). The increasing interest in innovation policy is reflected in growing demand for World Bank Group assistance from client countries in recent years.

Emerging economies and developing countries have increased their investments in R&D capability to enhance their innovation potential. For example, China had the fastest growth in R&D spending between 1999 and 2008, and Brazil, China, India, and the Russian Federation are now among the top R&D performers in the world (Dahlman 2014). Developing countries such as Kenya, Senegal, Tanzania, and Uganda are currently investing between 0.5 and 1 percent of their gross domestic product in R&D. Although investments in R&D themselves do not automatically lead to innovation, the recent focus on building up R&D capability is indicative of the changing geography of innovation potential. Several studies corroborate the impact of rising R&D investments on productivity, exports, and to some extent job creation, and poverty reduction (Lejour, Mervar, and Verweij 2008; Dercon and others 2009; IFPRI 2013).

Evaluation Rationale

The World Bank Group has supported lending and non-lending activities in science, technology, and innovation (STI) for more than three decades. But the contribution of the Bank Group's activities to innovation and entrepreneurship has never been subject to a comprehensive evaluation. Previous reviews of activities touching on this area have been limited to World Bank lending to support scientific and technological research, development, and capacity building (Goel and others 2003; Crawford and others 2006). This evaluation is the first comprehensive review to span the broad range of policies and instruments that the Bank Group institutions have used to support innovation and entrepreneurship.

The timing of this evaluation is significant for three reasons. First, innovation strategies, policies, and instruments are gaining in importance across all networks, sectors, and regions of the World Bank Group. The development of innovative ideas and tools is now considered essential for accelerating the pace of innovation and achieving development results (see Box 1.1). Second, an earlier, limited review of World Bank science and technology (S&T)

BOX 1.1 Recent Developments in the Bank Group Approach to Innovation

In 2011, the Innovation, Technology, and Entrepreneurship Practice (ITE) was established as one of the six global practices of the Finance and Private Sector Development Network. The practice's objective is to be a hub of networks and experts in the Bank Group and generator of knowledge around the topics of innovation, technology, and entrepreneurship. The ITE Practice supports innovation-driven productivity with the goal of creating more and better-paying jobs, increase success rates of growth-oriented small and medium-size enterprises and entrepreneurs, and foster inclusive innovation by strengthening private sector innovation and entrepreneurial capacity.

SOURCE: World Bank.

projects found no consistent Bank Group approach or strategy for catalyzing change in development (Crawford and others 2006). Third, lessons of experience can feed into the strategies and products of the new Finance and Private Sector Development Network (FPD) intended to foster innovation in addressing emerging development challenges, both within FPD and across the Bank Group.

This evaluation assesses how well the World Bank Group is fostering innovation and entrepreneurship in client countries. It is intended to serve both accountability and learning purposes. The goal is to inform future Bank Group strategic directions and enhance program and project implementation.

The evaluation objectives are as follows:

- Assess what specific interventions on innovation and entrepreneurship were expected to achieve and how they are addressed in strategies and project documents.

- Assess whether the expected results from these interventions were achieved.

- Find out why certain results occurred or did not occur as expected and draw lessons for the future design and implementation of strategies and operations.

Before examining the conceptual framework that is the foundation for this evaluation, it is important to understand the rationale for Bank Group support for innovation and entrepreneurship.

Conceptual Framework for the Evaluation

The literature provides different perspectives on how governments and development agencies can foster innovation and growth. One perspective links innovation to competition and market entry. Aghion (2006) argues that there is an inverted-U relationship between competition and innovation, implying some optimal degree of competition. In addition to competition and market entry, Aghion identifies investment in higher education, reform of credit markets, and management of the economic cycle as important ways of fostering innovation and growth. The OECD links trade openness to innovation because trade allows the movement of new technologies around the world. This mobility of technology increases the size of markets both for the innovator and for those that apply the innovation. Foreign direct investment (FDI) has also been associated with innovation processes. Girma, Gong, and Görg (2008) show that privately and collectively owned firms that participate in foreign capital and have good access to domestic banks innovate more than other firms do.

The conceptual framework for this evaluation is developed from theoretical literature relating to innovation processes. One strand in the literature emphasizes that the main rationale for development agencies, such as the World Bank Group, is market failures in the production of public goods, such as basic research. In the absence of countervailing institutions, particular types of market and government failures lead to the under-provision of certain types of research.[6] Other market failures lead to incentive problems, uncertainties, and information asymmetry that weaken the incentives for entrepreneurs to invest in innovative activities. In such cases, the correction of market and government failures provides the justification for Bank Group institutions' support for innovation and entrepreneurship.

Government failures are quite pervasive in innovation, including the failure to promote a policy environment that promotes innovation and fosters entrepreneurship, poor STI policies, and corruption (Dahlman 2014).[7] If these failures are successfully managed, then innovation and entrepreneurship are expected to increase. The implementation or commercial application of innovations will in turn promote higher productivity, competitiveness, and growth and will help reduce poverty.

More recent literature developed around concepts of national innovation systems has emphasized the systemic nature of innovation processes, in which innovation policy focuses on the flow of technology and information in a system of interaction among research institutions, universities, firms, and people (Freeman 1995; Lundvall 1992; Nelson 1993; Edquist 1997). In this perspective, support for innovation and entrepreneurship includes policy response to other bottlenecks and failures that impede innovation and entrepreneurial activities.[8] Four types of failures are included: capability failures, failures in institutions, network failures, and framework failures (Arnold 2004). Besides correcting market and government failures, innovation interventions also need to address aspects such as strengthening firms' capabilities to innovate and/or make more effective use of new technology, strengthening interactions among research institutions, universities, and firms as well work on the broader enabling environment in which innovation takes place (OECD and World Bank 2009; UNCTAD 2011). Thus, innovation is not conceptualized as a linear process but as a system that integrates production of knowledge and technology, its use by firms and other actors, and the interaction between producers and users of knowledge and technology.

These theoretical perspectives provide the building blocks for the conceptual framework that guides this evaluation, as shown in Figure 1.1. The inclusion of other bottlenecks or failures does not invalidate the correction of market and government failures as important policy rationale. Instead, combining policy responses to both types of failures in the conceptual

FIGURE 1.1 Conceptual Framework for Assessing World Bank Group Support for Innovation and Entrepreneurship

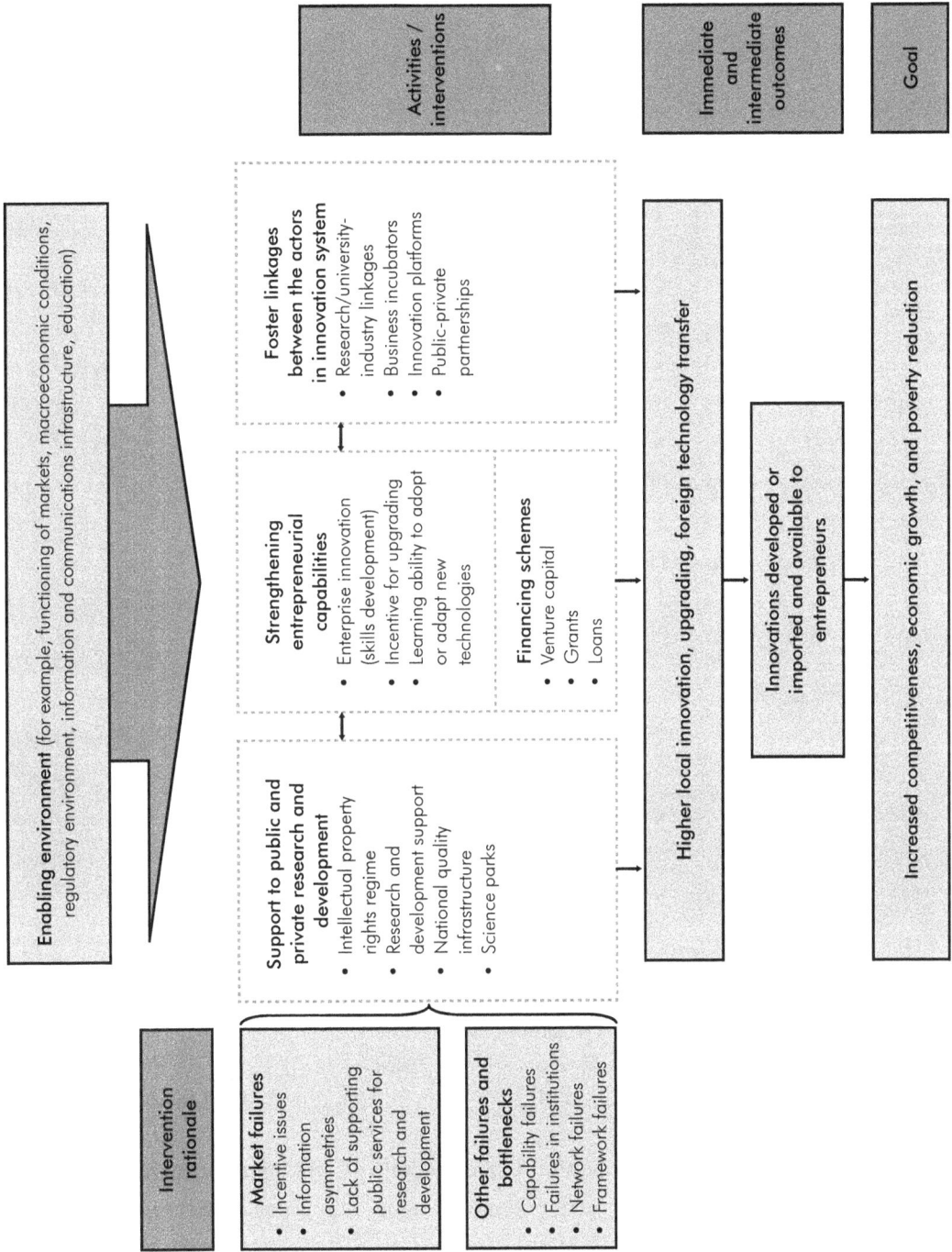

Enabling environment (for example, functioning of markets, macroeconomic conditions, regulatory environment, information and communications infrastructure, education)

Activities / interventions

Intervention rationale

Market failures
- Incentive issues
- Information asymmetries
- Lack of supporting public services for research and development

Other failures and bottlenecks
- Capability failures
- Failures in institutions
- Network failures
- Framework failures

Support to public and private research and development
- Intellectual property rights regime
- Research and development support
- National quality infrastructure
- Science parks

Strengthening entrepreneurial capabilities
- Enterprise innovation (skills development)
- Incentive for upgrading
- Learning ability to adopt or adapt new technologies

Financing schemes
- Venture capital
- Grants
- Loans

Foster linkages between the actors in innovation system
- Research/university-industry linkages
- Business incubators
- Innovation platforms
- Public-private partnerships

Immediate and intermediate outcomes

Higher local innovation, upgrading, foreign technology transfer

Innovations developed or imported and available to entrepreneurs

Goal

Increased competitiveness, economic growth, and poverty reduction framework allows IEG to identify the different types of policy rationales that have guided Bank Group support for innovation as well as the associated interventions.

As the middle of the figure shows, the conceptual framework identifies four types of targeted interventions that have been used to support innovation and entrepreneurship within a broad-based enabling environment. This framework recognizes that the process of innovation and diffusion of innovations take place within an economic environment and institutional context.[9]

The enabling environment for innovation also affects the extent to which firms have incentives and rewards to undertake innovations as well as the interest of foreign agents to contribute to domestic innovation efforts, the communication capabilities, and the feasibility of domestic agents to gain access to foreign sources of innovation. An exhaustive evaluation of the impact of the enabling environment is beyond the scope of this evaluation. However, the country case studies in Chapter 2 highlight how the enabling environment affects the performance of innovation in different country contexts. The review of corporate, sector, and country strategies also provides insights on how the enabling environment is addressed in strategic approaches to support innovation and entrepreneurship.

The four targeted interventions aim to directly foster innovation and entrepreneurship through support for public and private R&D, strengthening entrepreneurial capabilities, provision of financing schemes for entrepreneurs, and fostering linkages between various actors in the innovation system. These interventions aim to build countries' capabilities to innovate, as well as encourage firms to innovate and grow. In this evaluation IEG gives major emphasis to the design and results of targeted interventions in the project-level analysis in Chapters 3 and 4, as well as in lessons that are drawn in Chapter 5.

The conceptual framework allows for joint causality: it may be that promoting innovation and entrepreneurship will only succeed if a policy and regulatory environment exists, providing an environment conducive for entrepreneurs to create wealth through innovation. Therefore, some of the measures to deal with market and government failures, as well as other bottlenecks impeding innovation and entrepreneurship, may have a complementary impact.

The World Bank Group does not have a comprehensive strategy and results framework for projects that support innovation and entrepreneurship. This is partly because the agenda on innovation and entrepreneurship is still evolving. Different dimensions of that agenda are addressed in different sectors, whose activities may or may not be well integrated or aligned. Operational guidance says that the public sector should fill the gaps left by the private sector—but should not attempt to do what the private sector would do on its own. Assessing whether this support complements or replaces the private sector is a critical step in assessing the results achieved by Bank Group actions.

Evaluation Questions and Criteria

Using the conceptual model, IEG identified a set of evaluation questions to assess World Bank Group support for innovation and entrepreneurship. These questions span sector and regional strategies, project performance, and results. The overarching question is: To what extent did targeted Bank Group interventions foster innovation and entrepreneurship that were intended to transform new ideas into greater competitiveness, economic growth, and poverty reduction? The focus is not on innovations in donor delivery mechanisms but rather on support that fosters innovation in client countries and enterprises.

Specific evaluation questions are in three categories: the relevance and alignment of the Bank Group agenda on innovation and entrepreneurship, the effectiveness and efficiency of its interventions, and the results and learning agenda.

i.   Does the Bank Group provide adequate guidance to support the right interventions?

   a.   Do strategies provide adequate guidance and principles for selection and design of interventions?

   b.   What are the main rationales for interventions that support innovation and entrepreneurship?

   c.   What are the interventions and mechanisms used to support innovation and entrepreneurship?

ii.   Are interventions that support innovation and entrepreneurship effective and efficient in achieving their objectives?

   a.   To what extent do projects and project components achieve their stated objectives?

   b.   To what extent do projects and project components achieve their expected outcomes?

   c.   To what extent do interventions achieve these outcomes efficiently?

iii.   Is the Bank Group learning from its experiences in supporting innovation and entrepreneurship?

   a.   What factors are associated with what works or does not work, using both Bank Group and outside evidence?

   b.   Are mechanisms for sharing experiences, best practices, and learning within and across the Bank Group institutions working adequately?

   c.   What do these lessons imply for future Bank Group support for innovation and entrepreneurship?

Scope

The evaluation covers lending (investment) and non-lending activities that directly support innovation and entrepreneurship in Bank Group client countries. Following the conceptual framework, four main types of targeted interventions designed and implemented by the Bank Group are considered:

- *Support to R&D infrastructure:* This includes innovation that supports public services such as public research institutions, S&T parks, and public research universities; national quality infrastructure; and regulation including intellectual property rights (IPR). These activities are mainly funded by the World Bank.

- *Strengthening entrepreneurial capabilities:* These comprise building skills and management capabilities of entrepreneurs, with a view to improving their business performance. These interventions also involve efforts to facilitate firm expansion and growth via technological upgrading such as acquisition of new technology through technology transfer and/or diffusion; upgrading of existing products and processes; and introduction of new products, processes, and marketing methods.

- *Financing support for early-stage start-ups:* This consists of venture capital funds providing financing for new and existing innovative small and medium-size enterprises (SMEs).

- *Foster linkages between actors in the innovation system:* These interventions include strengthening interactions among public research institutions, universities, and industry to help bring innovations, in terms of new products and products, to market. Mechanisms such as matching grants and business incubators have been used to provide incentives for entrepreneurs to commercialize R&D products that have been developed in public research systems. These interventions are supported by the World Bank and the International Finance Corporation (IFC) through initiatives such as infoDev (see Box 3.1).

To maintain the focus of the evaluation, IEG did not cover every dimension that could conceivably impinge on innovation and entrepreneurship. Besides looking at country case studies and strategies that support innovation and entrepreneurship, IEG did not assess projects aiming to improve the enabling environment, which may have an impact on all kinds of private sector activities—such as support to basic or higher education or the overall business and regulatory environment.

The evaluation's focus on entrepreneurship is limited to Bank Group support for the development and growth of new enterprises, such as science parks and business incubators as well as existing firms that engage in innovative activities. Thus, it does not cover all entrepreneurship projects by the Bank and IFC that have focused on capacity building for SMEs. Such projects are only considered in cases where IEG had evidence that interventions focused on building the innovative capacity—to develop new products, processes, business models—of new firms or existing innovative SMEs. It also excludes support for management education or programs to support general business development, such as information services consulting. Also, because the new practice groups in FPD were established so recently, it is too early to evaluate their performance; thus, the activities of those groups were not evaluated.

The distribution of interventions across the World Bank Group that are covered in this evaluation is summarized in Table 1.1.

Rationale for Bank Group Intervention to Support Innovation and Entrepreneurship

World Bank Group client countries have requested support for innovation and entrepreneurship projects in pursuit of their development strategies to develop or maintain a competitive edge in global markets or diversify from resource-based to innovation-driven growth strategies. The rationale for World Bank Group interventions to support innovation and entrepreneurship rests on two claims: first, that innovation and entrepreneurship can be important for growth and poverty reduction and, second, that markets and governments often fail to create an enabling framework for innovation and entrepreneurship by private enterprises.

The World Bank Group's support for innovation and entrepreneurship has evolved. In the early years, the correction of market and government failures provided the major justification for supporting innovation and entrepreneurship projects (Appendix A). Bank Group interventions support governments and the private sector by helping provide solutions for market failures as well as government failures that restrict private investment in innovation that could generate economic and social benefits. Thus, the justification for World Bank (International Development Association and International Bank for Reconstruction and Development) interventions hinges on supporting corrective actions that reduce the gap between the value of what the private sector would provide alone and what would be socially desirable. Innovation may also help reduce poverty, irrespective of its impact on growth, adding to the rationale for public sector support.

TABLE 1.1 Targeted Interventions Supporting Innovation and Entrepreneurship across the Bank Group

Intervention	World Bank	IFC	MIGA
Support to public R&D			
R&D funding	X	X	
R&D capacity building	X		
National quality infrastructure	X		
Intellectual property rights	X		
Support for capacity building	X		
Strengthening entrepreneurial capabilities	X	X	
Skills development to SMEs/farmers	X	X	
Technological upgrading	X	X	
Financing schemes	X	X	X
Venture capital	X	X	
Loan/grants to SMEs	X	X	X
Fostering linkages between the actors in the innovation system			
Business incubators	X		
University-industry linkages	X		
Research-extension and farmer linkages	X		
Other linkages (Diaspora, among private sector)	X		

SOURCE: IEG.
NOTE: R&D = research and development; SME = small and medium-size enterprise.

WORLD BANK RATIONALE

The emphasis on correcting market and government failures focused early Bank support for innovation on projects that emphasized investments in public research infrastructure, improvement in efficiency of public sector R&D systems, and efforts to help the private sector commercialize products from R&D. Examples of projects approved in the 1990s that sought to achieve these objectives include the following:

- The Brazil Science and Technology Reform Project (1997), whose objective was to improve the overall performance of Brazil's S&T sector by undertaking activities that promoted scientific research and technological innovation in an efficient manner (project activities included support for S&T research and advanced training, strengthening IPR, and national quality infrastructure).

- The National Agricultural Technology Project in India (1998), whose objectives were to improve the efficiency of the Indian Council of Agricultural Research Organization and management system, enhance the performance and effectiveness of priority research programs and of scientists in responding to the needs of farmers, and develop models that improve the effectiveness and financial sustainability of the technology dissemination system with greater accountability to and participation by the farming communities.

Some projects focused on improving firm-level competitiveness. For example, the Industrial Technology Development Project in Indonesia (1995) sought to enhance the competitiveness of Indonesian industry, particularly by SMEs by providing public and private technology support services, facilitating access to public and private service providers, strengthening public technology support institutions, and improving formulation and coordination of industrial technologies.

In some projects the Bank used Learning and Innovation Loans to finance experimentation, learning, and piloting of promising STI initiatives prior to supporting large-scale interventions. For example, Chile's Millennium Science Initiative Project (1999) used a Learning and Innovation Loan to demonstrate significantly improved performance in a highly selected segment of the country's S&T system and thereby revitalized the country's S&T system. In the Nicaragua Competitiveness Learning and Innovation Loan (2000), funding was used to test public-private partnerships for developing consensus and introducing reform on business environment issues, and to pilot sustainable information technology-based business development services. An important feature of these projects is that the focus was on strengthening the supply of knowledge and technology through increased funding for public sector research and equipment, improving national quality infrastructure, and IPR. At the firm level, some projects provided support to stimulate R&D in the private sector through cooperative R&D activities between firms and the S&T community and assisted with upgrading the technological capabilities of firms.

A second generation of projects, approved in the 2000s, broadened the conceptualization of innovation beyond correction of market and government failures by addressing other bottlenecks or failures. For example, the Science for Technology Project in Chile (2003) had a major objective of supporting the development of an effective innovation system by establishing a strong and coherent policy framework, promoting high-quality and relevant S&T activities, and supporting key interfaces in the innovation system, especially between the public and private sectors as well as international linkages. The Croatia Science and Technology Project (2005) sought to strengthen and restructure selected R&D institutions to promote applied research, while maintaining their scientific excellence, and increase the ability of enterprises to develop, use, adapt, and commercialize technology. The objective of the India National Agricultural Innovation Project (2006) was to accelerate the collaborative development and application of agricultural innovation among public research organizations, farmers, the private sector, and other stakeholders.

More recent projects focus on strengthening sector and firm-level competitiveness, diversification from factor-driven to innovation-driven growth, and inclusive innovation. Bank support for innovation and entrepreneurship in these projects has—in addition to strengthening the infrastructure for S&T—increasingly considered how firms, farms, and other public sector actors use knowledge and technology, as well as considering the interactions between suppliers and users of technology and the enabling environment in which the process of innovation and development of entrepreneurship takes place.

IFC AND MIGA RATIONALE

Development institutions that focus on the private sector, such as IFC and the Multilateral Investment Guarantee Agency (MIGA), make investments expected to be additional to what the private sector itself would do and to support private firms in ways that help developing countries achieve sustainable economic growth (IFC 2011). IFC's interventions have supported important forms of innovative activities involving introduction of products, processes, and business models that are new to particular industries or firms. Its support for transfer of technology and know-how has helped increase the innovative capacities of firms and introduce important sources of innovation.

Among other things, clients have benefitted from IFC's involvement in projects through its provision of long-term finance that cannot be provided by commercial lenders; its stamp of approval and the comfort that it provides to other investors and its role as an honest broker in facilitating negotiations among different parties; and the introduction of best practices from its global experience. Many local banks are not willing to provide financing for new projects, for projects introducing new products that are untested in markets, or for start-ups that do not have significant assets for collateral because they are considered risky.

When markets fail, the private sector, on its own, under-provides investments for goods or services that are socially desirable. Financing for an IFC project would be justified if it brought an innovation to the local market (such as a new kind of mortgage lending) that would not have existed without the intervention. For example, IFC's participation in the first mortgage-backed securitization by a Russian bank provided a seal of approval that was critical in the development of the secondary mortgage market in the country. The demonstration effect from introducing this new mortgage product helped develop the residential housing sector by providing a refinancing mechanism that was replicable and used by other local banks. In Guatemala, IFC's support for the first geothermal plant provided a positive signal to other equity investors; IFC also acted as an honest broker in negotiations between the state utility company and a group of partner companies investing in the project.

In other cases, an intervention by IFC or MIGA would be justified if it demonstrated the financial feasibility of an innovation that was previously uncertain or that was not thought to be financially feasible in the local market. For example, IFC provided long-term financing for a number of companies in the ICT sector in a number of countries, such as Albania, Chad, the Democratic Republic of the Congo, the Dominican Republic, and Sierra Leone, that were perceived to be high-risk countries. Through its involvement, IFC helped transfer technology and know-how, demonstrated the commercial viability of these investments, and highlighted the potential for the sector, sending a positive signal to other investors.

In financial markets, IFC and MIGA helped introduce new banking products and established best practices in microfinance in countries perceived to be high risk, such as Afghanistan, the Democratic Republic of the Congo, and Sierra Leone. IFC has also supported venture capital financing for early-stage start-ups in developing countries, providing equity and quasi-equity financing as well as training and management support that local banks would provide for high-risk projects. To the extent that IFC and MIGA's involvement results in a net increase in innovation, the institutions' intervention would improve national welfare.

WORLD BANK GROUP ANALYTICAL AND ADVISORY SUPPORT

In addition to addressing market and government failures that limit private investments in innovation, the Bank Group institutions support innovation in government advisory roles. Their support seeks to improve the operation of the government in providing services and public goods that the private sector does not provide. This includes projects that help governments enable innovation that supports IPR. The rationale for this activity is rooted in the expertise of the Bank and IFC and their ability to mitigate public sector failures.

The World Bank has supported research and applied analytic work, workshops, and policy reviews, targeted technical assistance, and investment projects (loans and investments to develop innovation capacity, such as in technical and higher education and research infrastructure). The Bank Group has also funded projects that use innovative processes or deliver innovative goods and services or investment guarantees that facilitate such investments by the private sector. The relative importance of each varies among the three Bank Group institutions. The World Bank has the largest range, followed by IFC, whose strength is in fostering innovation through FDI in addition to its own input into projects.

OTHER INSTITUTIONS' WORK

Other multilateral financial institutions also have been supporting STI and entrepreneurship in developing countries. The Inter-American Development Bank and the European Bank for Reconstruction and Development have a roughly similar range of activities (although they do not have the investment guarantee function of MIGA). The Inter-American Development Bank also has an active lending, research, and policy advisory program on innovation (IDB 2011).

In addition to their innovation loan and investment projects, these institutions enable important policy discussions and dissemination of ideas through workshops and publications. Among them, the World Bank probably has the largest policy role, as it invests the most in policy advice—not only as part of the preparation of regular loan projects, but also through its research and policy work unrelated to loan projects.

Evaluation Methodology

The approach for this evaluation is non-experimental and combines qualitative and quantitative methods to answer the evaluation questions. All the evaluation questions are descriptive or normative, with performance assessed against criteria for relevance, effectiveness, and efficiency. Evidence for the evaluation comes from a portfolio review of World Bank, IFC, and MIGA projects; semistructured interviews with managers, task team leaders, transaction managers, and portfolio staff in all three institutions; field surveys of clients, beneficiaries, and key stakeholders in two World Bank–supported and two IFC-supported projects; and desk reviews of World Bank, IFC, and MIGA project appraisal, supervision, and evaluation documents. The evaluation also draws extensively from an external evaluation of an innovation program supported by the World Bank (Dahlman 2014). Additional information was obtained from a literature review, secondary data sources, and background materials. Appendix B has detailed information on specific methodologies and examples of evaluation instruments.

For the portfolio review, IEG used World Bank, IFC, and MIGA project databases between FY00 and FY11 to identify both closed and active projects focused on innovation and entrepreneurship. World Bank sector and theme codes, however, do not use innovation, entrepreneurship, or related terms to report on Bank activities. Nor does IFC or MIGA have a system that officially records or tracks innovation. Thus, IEG adopted an alternative approach to identify relevant projects and activities (Appendix B). The investment portfolio considered in the evaluation included 119 Bank projects; 300 IFC investment projects; and 108 MIGA projects. In addition, IEG reviewed 268 World Bank advisory and analytical activities (AAA) and projects and 84 IFC Advisory Service projects.

The report is in five chapters. Chapter 2 presents findings from country experiences with innovation strategies and reviews how World Bank Group strategies have addressed innovation and entrepreneurship. Chapter 3 explores issues related to project design, including the characteristics of innovation and entrepreneurship projects across the Bank Group and the different types of interventions that the Bank Group has used to support innovation and entrepreneurship. Chapter 4 assesses the performance of projects and interventions supporting innovation and entrepreneurship based on IEG's evaluative evidence. Lessons learned from project examples and mechanisms for sharing knowledge and learning within sectors and networks as well as across the Bank Group are examined in Chapter 5. The evaluation concludes with a synthesis of findings and recommendations.

Endnotes

[1] Product innovation involves the introduction of a new or significantly improved good or service; process innovation, the implementation of new or significantly improved production or delivery method; marketing innovation, the implementation of new marketing methods that involve substantial changes in product design or packaging, product placement, product promotion, or pricing; and organizational innovation, the implementation of new organizational methods in a firm's business practices, workplace organization, or external relations (OECD 2005).

[2] Baumol (2010) distinguishes between the innovative entrepreneur and replicative entrepreneur. An innovative entrepreneur is a firm that comes up with new ideas and puts them into practice; a replicative entrepreneur can be anyone who launches a new business venture, regardless of whether similar ventures already exist.

[3] Iacovone and Qasim (2013) discuss a comprehensive framework to promote entrepreneurship that involves analysis of firm entry, growth, and exit.

[4] Innovation is also important for developed countries to sustain growth. Recently, it has been argued that the policy agenda for promoting innovation-led growth for high-income and middle-income countries is converging. Although most developed countries have articulated and adopted innovation strategies, dynamic middle-income countries, such as Chile, China, Mexico, and South Africa, are giving greater priority to promote more innovation-driven growth (OECD and World Bank 2009).

[5] There is no overall agreement on the definition of this term. It is generally used to refer to innovations that are focused on goods and services relevant to the needs of people at the base of the economic pyramid.

[6] Nelson (1959) discusses the externality and incentive problems that provide a rationale for public policy support for efficient basic research.

[7] Government failures specifically related to innovation and entrepreneurship also include common challenges such as lack of coordination across various ministries and agencies involved in formulation and implementation of policies related to innovation and entrepreneurship; limited stakeholder consultation and participation; and lack of metrics for innovation performance. Bank interventions, especially technical assistance, are often designed to address one or several of these shortcomings.

[8] This perspective, emphasizing the inadequacies of concepts of market and government failures as a guide to policy and systemic nature of innovation processes, draws from more recent theoretical developments in evolutionary economics (see UNCTAD 2011).

[9] This is what the OECD calls the "framework conditions."

References

Aghion, Philippe. 2006. "A Primer on Innovation and Growth." Bruegel Policy Brief. Issue 2006/06.

Arnold, Erik. 2004. "Evaluating Research and Innovation Policy: A Systems World Needs Systems Evaluation." *Research Evaluation* 13(1): 3–17.

Baumol, William J. 2010. *The Micro-theory of Innovative Entrepreneurship*. Princeton, NJ: Princeton University Press.

———. 2002. *The Free-Market Innovation Machine: Analyzing the Growth Miracle of Capitalism*. Princeton, NJ: Princeton University Press.

Crawford, Michael, C. César Yammal, Hongyi Yang, and Rebecca L. Brezenoff. 2006. *Review of World Bank Lending for Science and Technology, 1980–2004*. Washington, DC: World Bank.

Dahlman, Carl. 2014. "Innovation and Entrepreneurship: Framework, Lessons from International Experience, and Implications for the World Bank Group." Independent Evaluation Group Background Paper, Washington, DC.

Dercon, S., Daniel O. Gilligan, John Hoddinott, and Tassew Woldehanna. 2009. "The Impact of Agricultural Extension and Roads on Poverty and Consumption Growth in Fifteen Ethiopian Villages." *American Journal of Agricultural Economy* 91(4): 1007–21, doi:10.1111/j.1467-8276.2009.01325.

Dutz, Michael. 2011. "Competition and Innovation-Driven Inclusive Growth." Policy Research Working Paper 5852, World Bank, Washington, DC.

———. 2007. *Unleashing India's Innovation: Towards Sustainable and Inclusive Growth*. Washington, DC: World Bank.

Edquist, C. 1997. "Systems of Innovation Approaches—Their Emergence and Characteristics." In C. Edquist (ed.), *Systems of Innovation: Technologies, Institutions and Organizations*. London: Pinter/Cassell.

Freeman, C. 1995. "The National System of Innovation in Historical-Perspective." *Cambridge Journal of Economics* 19(1): 5–24.

Girma, Sourafel, Yundan Gong, and Holger Görg. 2008. "Foreign Direct Investment, Access to Finance, and Innovation Activity in Chinese Enterprises." Kiel Institute for the World Economy.

Goel, Vinod K., Ekatarina Koryukin, Mohini Bhatia, and Pryanka Agarwal. 2003. *Innovation Systems: World Bank Support of Science and Technology Development.* Washington, DC: World Bank.

Iacovone, Leonardo, and Qursum Qasim. 2013. "Entrepreneurship Policy Brief: An Introduction for Analysis and Promotion." World Bank, Washington, DC.

IDB (Inter-American Development Bank). 2011. *The Imperative of Innovation: Creating Prosperity in Latin America and the Caribbean.* Washington, DC: IDB.

IFC (International Finance Corporation). 2011. *International Finance Institutions and Development Through the Private Sector.* Washington, DC: World Bank.

IFPRI (International Food Policy Research Institute). 2013. *Global Food Policy Report.* Washington, DC: IFPRI.

Lejour, A.M., Andrea Mervar, and Gerard Verweij. 2008. "The Economic Effects of the Lisbon Agenda Targets: The Case of Croatia." Background paper for the Croatia EU Convergence Report 2009, No. 48879-HR, SSRN Working Paper.

Lundvall, B-A. 1992. *National Systems of Innovation: Towards a Theory of Innovation and Interactive Learning.* London: Pinter.

Nelson, Richard R. 1959. "The Simple Economics of Basic Scientific-Research." *Journal of Political Economy* 67(3): 297–306.

Nelson, Richard R., ed. 1993. *National Innovation Systems.* New York: Oxford University.

OECD (Organisation for Economic Co-operation and Development). 2011. *Measuring Innovation: A New Perspective.* Paris: OECD.

———. 2008. *A Framework for Addressing and Measuring Entrepreneurship.* Paris: OECD.

———. 2005. *Frascati Manual: Proposed Standard Practice for Surveys on Research and Experimental Development,* 6th ed. Paris: OECD.

OECD (Organization for Economic Co-operation and Development) and Eurostat (Statistical Development Office of the European Communities). 2005. *Oslo Manual: Guidelines for Collecting and Interpreting Innovation Data.* Paris: OECD.

OECD and World Bank. 2009. *Innovation and Growth: Chasing a Moving Frontier.* Paris: OECD.

Pasquier, B., and Andrew Stone. 2008. *Fostering Entrepreneurship and Private Enterprise in the Arab Economies: The Path to Youth Employment and Career Development.* Washington DC: World Bank.

Solow, R. M. 1957. *Technical Change and the Aggregate Production Function. The Review of Economics and Statistics.* Cambridge, MA: MIT Press.

Srinivasan, T. N. 2004. *Annual Bank Conference on Development Economics: Accelerating Development.* Washington DC: World Bank.

UNCTAD. 2011. *A Framework for Science, Technology, and Innovation Policy Reviews.* New York: UNCTAD.

World Bank. 2010a. *Competition and Innovation-Driven Inclusive Growth.* Washington, DC: World Bank.

———. 2010b. *Innovation Policy: A Guide for Developing Countries.* Washington, DC: World Bank.

2

Strategies to Support Innovation and Entrepreneurship

CHAPTER HIGHLIGHTS

- There are large divides in innovation input and outputs across country income categories; developing countries seeking to catch up with successful middle-income and high-income countries must build their capacity to innovate.

- The Bank Group can play a vital role in helping build countries' innovation capacities, enabling them to acquire, adapt, and use innovations that have been developed elsewhere.

- Bank Group support for innovation and entrepreneurship in client countries can be enhanced in several ways, including support for technical and entrepreneurial capabilities and knowledge exchange.

- Bank Group strategic documents have signaled support for innovation and entrepreneurship, but they have not articulated an overarching vision for policy action.

- Myriad activities support innovation and entrepreneurship across the Bank Group, but a well-coordinated cross-sectoral set of actions has not yet emerged.

Experience from country perspectives helps illustrate the dimensions underscored by the conceptual framework developed for this evaluation. Countries at different levels of development increasingly recognize that innovation is critical for maintaining a competitive edge in the global economy, as well as for facilitating economic diversification and economic progress (OECD and World Bank 2009; OECD and IDRC 2010; UNCTAD 2011). These countries are seeking more effective ways to translate scientific and technological knowledge into new products, processes, and business models that foster innovation-driven growth. Some have requested support from the OECD and UNCTAD for in-depth review of innovation policy in order to diagnose their innovation systems and identify policy priorities to enhance their innovation performance (OECD and World Bank 2009; UNCTAD 2011).[1]

This chapter examines strategic approaches to innovation and entrepreneurship within the Bank Group, looking at trends and drivers in innovation processes in a variety of country development contexts. The chapter also assesses the extent to which key issues relating to innovation and entrepreneurship are addressed in Bank Group strategies. It concludes by identifying some principles that can be used to formulate policy priorities that support innovation and entrepreneurship in Bank client countries and that have been reflected in the conceptual framework of the evaluation.

The analysis of trends and drivers of innovation processes is based on a background paper that includes country case studies covering 10 countries at different stages of development.[2] The assessment of Bank group strategies is based on a desk review of recent Bank Group corporate, sector, regional, and country strategies over the past decade or more.

Client countries are looking to the World Bank Group to help them develop strategies and design policies and programs to facilitate innovation-driven growth that strengthens competitiveness (Yusuf 2009; Devan 2012). Some governments, with support from the Bank Group, are also pursuing strategies and initiatives to address inclusive innovation and environmental sustainability (Boxes 2.1 and 2.2).

Environmental sustainability issues, particularly innovative approaches to deal with climate change, have featured prominently in some country policy dialogue with the Bank Group. China's Country Assistance Strategy (CAS), for example, focused on enhancing innovation and promoting development of the environment and energy efficiency, including new energy and new energy vehicles. The Bank Group has helped clients create incentives and mobilize resources that have supported the development, demonstration, transfer, and diffusion of climate change–related innovations. Box 2.2 provides examples of some of these initiatives.

In many emerging economies and developing countries, innovation frequently involves acquiring and making effective use of technologies that already exist and that are in widespread use elsewhere but may be new to the firm or the market, or used in new ways. Measures of innovation show that, on average, innovation linkages—the productive interaction among firms, the public sector, universities, and society—in most low- and middle-income countries are weaker than those in high-income countries.[3] INSEAD's innovation measures show that innovation outputs and inputs are strongly correlated with income levels (Figure 2.1). On average, innovation performance is stronger in high-income countries than in middle- and low-income countries.[4] Data from the World Bank Group Enterprise Survey also show strong linkages between country income levels and selected correlates of innovation (Figure 2.2).[5] There are large divides in innovation across geographic regions, with average performance in the more dynamic upper-middle-income countries in Southeast Asia such as China and Malaysia much higher than those in Africa, South Asia, and Latin America.

Bank Group initiatives supporting innovative climate change initiatives include:

- Participation in the Clean Technology Fund, a multidonor trust fund created as part of the Climate Investment Funds to provide scaled-up financing for the demonstration, deployment, and transfer of low-emission technologies that have significant potential for long-term greenhouse gas emissions savings

- Financing such as the Forest Carbon Partnership Facility, climate risk management products, and "Green Bonds"

- IFC's Climate Business Group, targeting innovative investments including technology transfer and opportunities for SMEs

- Work with the Global Environment Facility in financing several initiatives, including projects that focus on clean energy technologies.

SOURCE: IEG.

FIGURE 2.1 Average Scores on Innovation Inputs and Outputs, by Country Income Group

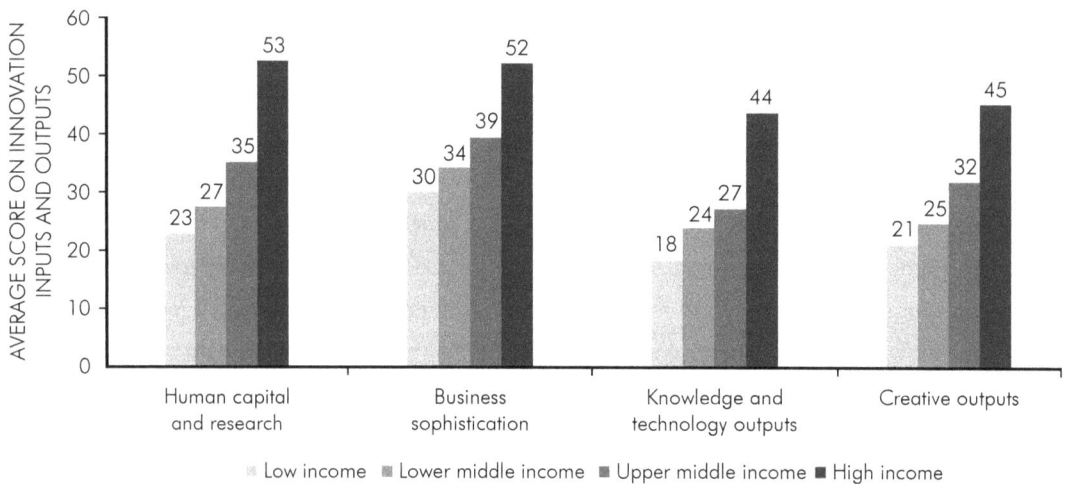

SOURCE: Global Innovation Index.
NOTE: Of the 141 countries in the index, 21 are low-income countries, 36 are lower-middle-income, 40 are upper-middle-income, and 44 are high-income.

FIGURE 2.2 Percentage of Firms Engaging Innovation Activities, by Country Income Group

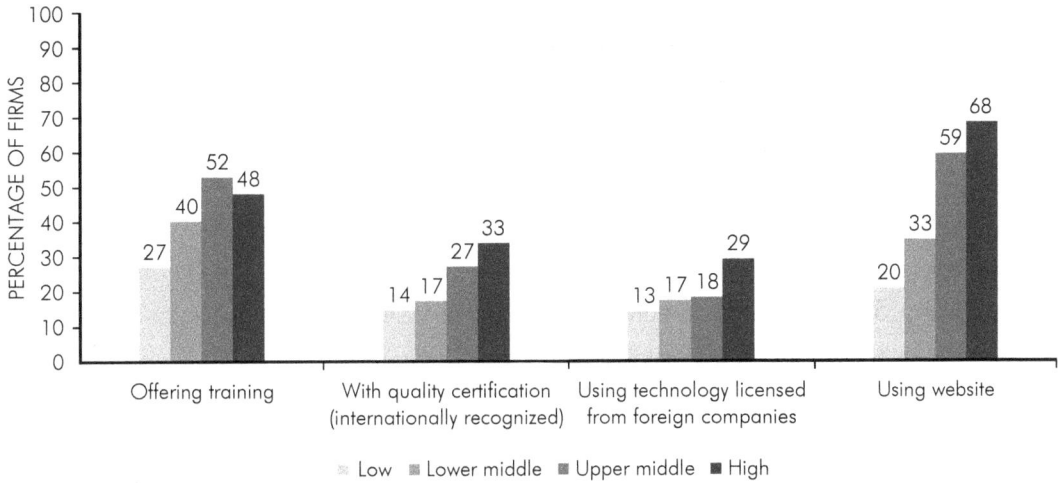

SOURCE: World Bank Group Global Enterprise Surveys, 2006–10.
NOTE: This covers 106 counties and 46,556 enterprises.

Building Innovation Capacity at the Country Level

Innovations may come from knowledge and technologies from foreign sources, from other users in the economy, or been created by domestic research from public institutions, universities, and private firms. However, increasingly many innovations are emerging from developing countries, including incremental and frugal innovation that has led to the redesign of products and business models that significantly reduce costs.[6] The strengthening of innovation capacity has been an important factor in countries that have experienced rapid and sustained economic growth. Emerging economies and developing countries seeking to pursue development strategies that foster growth must build the capacity to acquire, disseminate, and use technologies to promote innovation and encourage new and existing firms to invest in business opportunities.

The Bank Group can play a vital role in helping countries build their innovation capacity. There is no unique path to innovation that drives development, but experience shows that countries have used a variety of strategies to foster innovation and entrepreneurship.[7] The 10 country case studies done as background for this evaluation emphasize several elements as important drivers of innovation performance in these countries—the enabling environment, including a modern information and communication infrastructure; support for R&D and human capital development; entrepreneurial capabilities and linkages to tap into global knowledge; and financing arrangements (Appendix C). All these elements are directly linked to the context and targeted interventions identified in the conceptual framework for this evaluation.

The development and transfer of scientific results and inventions and their application to address development challenges or improve social welfare is less effective if the environment is not enabling. Similarly, entrepreneurs need an environment that is conducive to investment in innovative activities that develop new ventures and create jobs.

Countries with strong innovation performance demonstrate that getting the policy and institutional environment right is very important to ensure that policy measures (such as STI and entrepreneurial incentives) targeted to boost innovative activity and foster entrepreneurship are successful. A supportive policy and institutional environment are critical both for the incentive to innovate and for the allocation of resources. Singapore's excellent investment climate and supportive regulatory environment have played a key role in the country's successful innovation-driven growth, creating striking levels of innovation outputs and one of the most competitive economies in the world. High levels of investment in human resources or R&D are not sufficient if they are not allocated to activities that improve competitiveness.

Russia provides an example of the importance of the policy and institutional environment for the incentive to innovate and allocation of resources. Significant investments in human capital and R&D in Russia have not translated into a dynamic innovation system or competitive economy. Even though there have been significant political and economic development over the past decade, growth in the economy is driven by oil and gas exports. Investment in other sectors is unattractive and the competitiveness of Russian enterprises in the global economy is eroding in key sectors such as manufacturing. Recently attention is being given to reforms and initiatives that will help develop an economic and institutional environment that encourages investment and risk taking—and rewards such efforts.

As part of the enabling environment, ICT offers many opportunities to help build entrepreneurial capabilities by facilitating technology uptake and firm competitiveness. Rapid advances in ICT have helped reduce transaction costs and coordinate economic and social activities. For example, ICT facilitates easier access to global knowledge for scientists and innovators from developing countries; it also expands markets for entrepreneurs. These investments have therefore provided a complementary infrastructure for an effective knowledge-driven and entrepreneurial economy. In doing so, they facilitate linkages between various actors in the domestic and global innovation system.

An effective information and communication infrastructure has been an important factor for countries with the most successful innovation strategies. Most of them, like China, the Republic of Korea, and Singapore, have also become major producers of ICT hardware because that sector offered many opportunities for production and trade. In addition, the ICT sectors are still undergoing rapid technical change and are at the forefront of innovation in many areas, ranging from automation and process control to social media. Many countries are increasing R&D in ICT because it still offers many innovation opportunities. Some developing economies with low per capita incomes have been very effective users of the innovation potential offered by these generic technologies and have been leaders in their innovative applications, as in the successful example of mobile money, M-Pesa in Kenya.[8]

SUPPORT FOR R&D AND HUMAN CAPITAL DEVELOPMENT

To acquire and use knowledge, developing countries need R&D capability as well as education and skills. The mix of education and skills may be different, but some R&D capability is necessary to follow and obtain knowledge from abroad and adapt it to local conditions. Furthermore, domestic capability is necessary to benefit from the knowledge spillovers from FDI. Singapore is a relatively small economy, but it has made significant efforts to develop strong STI capabilities that permit it to take advantage of its intense interaction with the global innovation system. Countries like Kenya and Rwanda have made some progress, such as the innovative application of mobile money in Kenya and the use of mobile phones to provide up-to-date market pricing information to farmers, consumers, and traders in Rwanda. But these countries still have incipient STI capabilities, which make it more difficult for them to benefit from the global system. China, in contrast, has strong domestic capability; combined with a strategic government, it has been able to exploit global knowledge through formal means as well as by copying and reverse engineering.

Investment in basic, secondary, and tertiary education and skills development is critical for building innovation capacity in developing countries. India succeeded in ICT-enabled exports because it has a critical mass of educated and trained engineers. However, the low education and poor skills of its labor force have been major constraints to developing a competitive labor-intensive manufacturing export strategy. Investments in human capital and research need improvement to achieve higher levels of growth and innovation in India.

Good secondary and tertiary education is necessary for citizens to be able to absorb global knowledge and to move the value chain to more productive technologies, as in China. In addition, well-trained scientists and engineers are the basic input into more sophisticated R&D activities. China is investing massively in tertiary education, particularly for scientists and engineers, in pursuit of its aim to be a major innovative power. The need to invest in high-level human capital to improve innovation capabilities is a lesson that Brazil and Chile have also learned.

STRENGTHENING ENTREPRENEURIAL CAPABILITIES AND LINKAGES BY TAPPING INTO GLOBAL KNOWLEDGE

The direction and extent of innovation are shaped, in part, by learning processes that take place within and between firms (Arnold and Bell 2001; Bell 2007). Tapping global knowledge through various forms has been critical in many countries that have successfully developed innovation-driven and entrepreneurial economies. There are variations on how countries use global knowledge and technology as a strategy to foster innovation. Singapore, for example, continues to rely very heavily on FDI and global expertise. In contrast, Korea's strategy was to tap global knowledge but maintain local control. Thus, Korea limited FDI and opted instead to develop local knowledge through trade (capital goods, technical assistance, technology licensing, copying and reverse engineering), foreign education, and training, attracting its technical Diaspora back and investing heavily in R&D. China has used all these strategies, but it has also been very effective at tapping FDI.

China and India have benefitted from the spread of ideas through the Diaspora. In China, the technology industry is dominated by the Chinese Diaspora. The Indian Diaspora also maintains strong linkages with their home country. For example, Indian computer scientists in Bangalore constantly exchange ideas with their colleagues in Silicon Valley.[9] Both governments and firms benefit when they develop networks and skills by tapping into the global market through the Diaspora.

FINANCING SCHEMES

Countries have used different financing schemes to support new firms and innovative entrepreneurs. To address many capital market failures and challenges facing small innovative firms, governments have provided angel capital and given venture capitalists some special tax breaks or other benefits such as guarantees to help offset the extra costs and higher risks involved with innovative activities. China and Malaysia have provided special tax breaks to the first or pioneer firm to produce a new product, process, service, or form of organization. India and Korea have adopted the concept of the Small Business Innovation Research, a public-private partnership initiative developed in the United States, to provide incentives for entrepreneurs to develop new products, processes, and services.

In sum, this analysis of country experiences provides some insights about innovation strategies in specific development contexts. It does show that innovation strategies must be diverse, because countries differ in their resource endowments, challenges, and needs. Local market and policy realities help shape robust innovation strategies and foster entrepreneurship. Within this diversity of experience, five common principles emerge that can be useful in developing a framework to formulate more effective policies and programs that promote innovation and entrepreneurship:

- Support public investment in R&D that focuses on improving efficiency and relevance to end users as well as strengthening the use of research results in public policy decisions.

- Build domestic STI capabilities to make effective use of global knowledge. Education and skills depend on investment in basic, secondary, and tertiary education and other chances to learn, as well as investments in S&T. Such investments are integral to building an adequate enabling environment.

- Strengthen linkages between public R&D and private sector users of knowledge and technology and knowledge flows to the private sector. Incentives and other cost-effective mechanisms are needed to diffuse knowledge from R&D to firms and other end users. This includes entrepreneurial capabilities to leverage global knowledge and technology. In virtually all successful cases, tapping global knowledge was critical. Key channels for doing this include FDI, trade, foreign education and training, and attracting skilled and knowledgeable expatriates back to their country.

- Build a strong enabling environment, including effective use of ICT in a wide range of applications to foster innovation and benefit from the many business opportunities that ICT offers and significant investments in education and skill development.

- Provide flexible financing arrangements to encourage innovative firms to undertake risks in developing new products, processes, and services.

Each of these principles relates to the major intervention areas identified in the conceptual framework (see Figure 1.1). The next three chapters focus on targeted interventions and mechanisms that the Bank Group has used to address these issues, their performance, and the lessons that can be drawn from case studies and project evaluation. In the meantime, it is worth emphasizing that experience shows that countries are well advised to get the broad-based enabling environment right, including the effective use of ICT. Creating a supportive policy and institutional environment is fundamental to incentives, rewards, and risk taking, as well as innovation performance and success of policies and programs that foster innovation and entrepreneurship. In this sense, the enabling environment directly or indirectly affects all the targeted interventions.

Improving the relevance and efficiency of research systems and strengthening domestic STI capabilities are critical in supporting public and private R&D. Firms are at the center of innovation processes; thus, strengthening knowledge flows to the private sector, including those from global sources, is helping build entrepreneurial capabilities. Incentives include financing schemes and other forms of support that help firms address financial and capital constraints that enable them to innovate. Strengthening knowledge flows is also crucial in fostering linkages among the actors in an innovation system.

Innovation and Entrepreneurship in Corporate, Sector, Regional, and Country Strategies of the World Bank Group

Bank Group client countries are increasingly requesting support to help them develop innovation strategies as well as design policies and programs that would enhance innovation as an engine of growth. Issues relating to innovation and entrepreneurship have been addressed in various Bank Group strategies and policy documents. This section reviews how the issues have been addressed in corporate, sector, regional, and country strategies to offer a sense of the extent to which strategies address country needs in these areas. Besides corporate strategies and select regional and country strategies, the analysis focuses on strategies in four sectors where the bulk of the Bank Group work on innovation and entrepreneurship are concentrated: agriculture and rural development (ARD), private sector development (PSD), education, and ICT. PSD and ICT strategies are joint World Bank-IFC strategies, reflecting strategic directions in the two institutions; recent strategies and action plans in education and ARD involve collaboration across the World Bank Group.

CORPORATE STRATEGIES

At the corporate level, two strategic documents developed during the period covered by this evaluation provided some perspective on the evolution of thinking on innovation within the Bank Group. The 2001 Strategic Framework acknowledged the growing importance of the private sector and the development potential inherent in rapid technological advances. The document provided a framework for selectivity but left sector and country strategies to make the hard choices. More recently, the 2010 Post-Crisis Directions Strategy identified fostering innovation and competitiveness as an important policy action that will guide Bank Group efforts to achieve the strategic priority to create opportunities for growth. The strategy highlights several elements of the enabling environment, such as a robust investment climate, competition, policies that create a stable and sound financial sector, and promoting foreign investment, as critical requirements to encourage innovation, productivity, and a vibrant entrepreneurial private sector.

IEG's review of four annual strategic documents prepared by IFC—the 2000 and 2001 Strategic Directions and 2011 and 2012 Road Maps—found that the early documents barely acknowledged the importance of innovation in IFC's agenda. The later documents discussed the importance of PSD as a prerequisite for innovation, but discussion of specific support for innovation was mostly limited to climate change, with mention of a plan to include innovation as a priority in a forthcoming middle-income countries strategy.

Of the three MIGA Strategic Directions papers reviewed (FY05–08, FY09–11, and FY12–14), the FY09–11 strategy paper paid the greatest attention to innovation. But its focus is more on innovations in MIGA's product offering to its clients than on ways to support innovation at the corporate and country level.

The growing attention to innovation across Bank Group corporate documents, as well as IFC and MIGA strategic directions, seems to signal to staff and development partners that the innovation agenda is increasingly important. But these strategic documents have not articulated an overarching vision for innovation and entrepreneurship. There is a focus on improving the enabling environment mainly within the context of better public policy and on supporting institutions for private sector growth. Less attention is given to policy measures targeted to foster innovation and entrepreneurship. The lack of a corporate-level strategic direction on the role of innovation in the development process and limited articulation of innovation policies provides little guidance to staff on how to incorporate innovation and entrepreneurship activities more broadly in sector strategies and other work in the Bank Group.

SECTOR STRATEGIES

Sector strategies provide a conceptual framework for the Bank's work in each sector, an inventory of its experience, a shared understanding of sector priorities among anchor and regional staff, and a means to communicate strategic priorities with external partners (IEG 2012). The four sector strategy papers reviewed in this section emphasize innovation and entrepreneurship to varying degrees (Appendix C).

Early PSD strategies highlighted some dimensions of the enabling environment, but the more recent strategic action plan pays specific attention to the enabling environment for innovation and entrepreneurship, entrepreneurial capabilities, and financing schemes for existing firms and start-ups. In the 2002 strategy, innovation and/or entrepreneurship are not strategic objectives or expected outcomes that PSD activities aimed to achieve. However, one of the three strategic pillars—extending the reach of markets—addresses some elements of the enabling environment for innovation such as investment climate issues. The strategy also addresses direct public support, including financing to firms, particularly SMEs. But it does not pay attention to financing issues or challenges faced by start-ups or innovative entrepreneurs.

The 2009 PSD Mid-Cycle Implementation Progress Paper (World Bank 2009) identified mechanisms, such as matching grants, that can be used to help strengthen entrepreneurial capabilities in firms, but it did not identify actions to integrate them into its strategic priorities or programs. The absence of a clear innovation agenda in the PSD strategy has meant a lack of serious discussion about how activities undertaken in the sector can support innovation and entrepreneurship. Without such discussion, there is little chance for the articulation of principles and guidelines for PSD support in this area.

More recently, FPD has identified innovation as one of its four strategic pillars, and the new ITE Practice has made policy for innovation systems, technology transfer and diffusion, financing linkages for entrepreneurship, and inclusive innovation and green innovation key pillars in its 2012 Action Plan. These priority areas focus attention on key innovation policy interventions.

Education strategies have covered the enabling environment (as per the conceptual framework in Figure 1.1), support to public R&D, and strengthening entrepreneurial capabilities through skills development among the pillars of the conceptual framework. The 1999 Education Strategy acknowledged that rapid technological change and greater exposure to global competition implied a need for a more educated workforce that can innovate continuously. But the impact of sector activities on innovation was not articulated within a comprehensive vision for supporting capacity building in STI. The pillars to be pursued in the strategy—basic education for girls and in the poorest countries, early interventions, innovative delivery of education services, and education system reforms—might help achieve the implied innovation objective.

The 2011 Education Strategy focused on Learning for All, emphasizing the growing demand for technical and vocational education and training. One of its strategic priorities—strengthening educational systems—noted that a focus on tertiary education policy is necessary to promote STI. The strategy identified tertiary education in middle-income countries and skills development as a strategic priority. Supporting capacity building in STI is critical for building innovation capabilities and is a key component of public support for R&D. However, the focus in the current education strategy remained squarely on basic and secondary education and omitted any detailed discussion of how the Bank can support the building of STI capacity in client countries, including those in the low-income category.

ARD strategies have a long tradition of addressing public R&D in agriculture. More recently they emphasize a broad range of issues within the agricultural innovation system, including strengthening the linkages between technology development and other actors in agriculture innovation systems. The 2003 Rural Development Strategy gave substantial attention to innovation. Although it is not the primary emphasis, the strategy asserted that the Bank would support "sustainable intensification through the application of science" to improve agricultural productivity, continue to support agricultural innovation through the Consultative Group on International Agricultural Research, and help expand extension services so farmers could access new technologies. The focus has mainly been on support to public R&D systems, including S&T capacity building in agriculture at domestic, regional, and international levels.

Other interventions have focused on building the capabilities of farmer associations and fostering interactions among actors in the agricultural innovation system. The 2009 Agriculture Action Plan built on the consensus surrounding the 2008 World Development Report on agriculture and intensified the focus on innovation. The first pillar of the 2009 strategy focuses on raising productivity, which would be achieved largely through support for R&D-induced technology adoption to increase yields, the expansion of extension services, and scaled-up support for new technology generation with special emphasis on regional-specific approaches.

In addition to formal strategies, the agriculture and rural development sector published a sourcebook on agriculture innovation systems (World Bank 2012). The sourcebook addresses why investments in agricultural innovation systems are becoming important, as well as how specific approaches and practices can foster innovation in a wide range of contexts. It also provides detailed guidance on building, improving, and assessing country-based innovation systems.

The recent ICT strategy has put innovation at the core of its strategy, emphasizing entrepreneurial capabilities, financing for ICT entrepreneurs, and fostering linkages in innovation systems, such as in ICT-related business incubators. In addition, as a general purpose technology, ICT is an important component of the enabling environment for innovation and entrepreneurship (see Figure 1.1), but this aspect has not been emphasized in the ICT strategy. Innovation was not a strategic pillar in the 2002 ICT strategy, even though the document referred to organizational innovation as part of a successful implementation strategy to deliver its mandate.

This lack of emphasis on innovation changed with the 2012 strategy, which identified support for innovation as a strategic priority for the sector. "Innovation," one of three pillars of the 2012 strategy, aims to advance ICT to improve competitiveness and accelerate innovation and target ICT skills development. The strategy articulates a vision for the World Bank and IFC, working together to promote an enabling environment, strengthening entrepreneurial capabilities in information technology–related fields, financing that industry in emerging markets, and fostering linkages mainly through information technology–based business incubators.

This review of corporate and sector strategies shows that several dimensions of the enabling environment and targeted interventions for supporting innovation and entrepreneurship are being addressed in one way or another (see Table 2.1). However, these interventions are often designed and implemented within sectors with little or no coordination of activities or efforts to actively build on the comparative advantage of different sector teams to attain results at country levels. So far, the Bank Group has not articulated an integrated perspective for supporting innovation and entrepreneurship at the country level. A combination of actions across corporate and different sector strategic priorities is required in efforts to foster innovation and entrepreneurship in Bank Group client countries.

TABLE 2.1 Innovation and Entrepreneurship in Bank Group Strategies

Type of Strategy	Enabling Environment	Support for Public R&D Including Capacity Building	Entrepreneurial Capabilities	Financing Schemes	Fostering Linkages
Corporate	X				
Sector					
PSD	X		X	X	
ED	X	X	X		
ARD		X			X
ICT			X	X	X

SOURCE: IEG.
NOTE: ARD = agriculture and rural development; ED = education; ICT = information and communications technology; PSD = private sector development; R&D = research and development.

At the regional level, FPD regional staff worked in close collaboration with network staff to develop innovation strategies for Europe and Central Asia and the Middle East and North Africa Regions in the World Bank. The Europe and Central Asia strategy, the goal of which is to raise productivity and competitiveness, aims to align its innovation, technology, and entrepreneurship activities and identifies future opportunities for innovation work. The innovation strategy for the Middle East and North Africa Region identifies two priority areas for its interventions—innovations that have the highest impact on challenges in this region and areas where innovation makes a difference for inclusive and sustainable growth. At IFC, the regional strategy for the Latin America and the Caribbean Region includes support for competition and innovation.

These regional strategies are a good start in articulating a regional vision for innovation, and they demonstrate how the new FPD configuration can help link regional and sectoral strategies. However, much more needs to be done to identify innovation strategies for countries at different stages of development, as well as to provide a road map for how the regions would pursue the strategic priorities that have been identified.

Country demand for innovation and entrepreneurship is typically expressed in the Bank Group CASs. To get a sense of the demand for such support, IEG reviewed the CASs for 17 countries—6 lower income, 4 lower middle income, and 7 upper middle income—that had more than two innovation and entrepreneurship projects over the past decade.

The treatment of innovation and entrepreneurship in country strategies is varied. Upper-middle-income countries, such as Brazil, Chile, and China, give high priority to innovation in their development plans. Chile's CAS, for example, states that the government's agenda begins with innovation because the country lags behind fast-growing knowledge economies. In Brazil, innovation and productivity are a crucial part of the country's growth agenda. The country recognizes that innovation policy and support for entrepreneurship are critical for improving productivity and competitiveness.

CASs for lower-middle-income and low-income countries also prioritize innovation and entrepreneurship to improve competitiveness as well as diversification from resource-based to knowledge- or innovation-driven development. In Uganda, the national development strategy recognizes the role of STI in its growth strategy. The Mozambique CAS acknowledges that firms in the country will need to become increasingly competitive globally, emphasizing innovation and competitiveness to promote employment and exploit new sources of growth.

Bank Group support for innovation and entrepreneurship has responded to the needs and demands of client countries. In Chile, for example, the government requested or indicated interest in interventions related to innovation and entrepreneurship. In Brazil, the Bank increased its support in competition and innovation policy, areas where the government had defined a program of initiatives to improve growth potential and competitiveness. The CAS for Croatia noted the importance of adopting innovative technologies to improve productivity and achieve European Union accession. Bank support was requested for enhancing the IPR and R&D in the private sector.

Most of the Bank Group's development partners that are seeking to enhance competitiveness through innovation-driven growth recognize that an enabling environment (macro conditions, competition, business environment and regulations, intellectual property protection, and so forth) is critical. Armenia's CAS, for example, prioritizes Bank support for telecommunications and for establishing an innovation fund for centers of higher education; Chile strives to design and implement policies that improve competitiveness linked to ICT, research, and innovation; Mexico prioritizes improvements in investment climate, STI policy, and infrastructure; and China emphasizes improved firm competitiveness by removing barriers to competition and putting in place incentives for innovation.

Several CASs identify support for public and private R&D and capacity building as key elements of their innovation agenda. Uganda, for example, identifies investments in STI as a key intervention area in its growth agenda. Kazakhstan intends to promote innovation through support for R&D investments and tertiary education. Bangladesh also prioritizes investment in S&T in its sector strategies to improve productivity and efficiency.

Countries such as Armenia, Brazil, Colombia, Kazakhstan, Mexico, Mozambique, and Peru are requesting Bank Group support to strengthen entrepreneurial capabilities. In particular, IFC support is requested to foster technology upgrading through technology transfer and diffusion, South-South knowledge transfers, and access to new markets and products. Brazil requested IFC support for innovation and entrepreneurship by promoting South-South knowledge transfers and encouraging access to new markets and products. Other countries such as Colombia and Kazakhstan requested IFC's support in fostering entrepreneurship or developing innovative business models as part of their modernization agenda.

The Bank Group also seeks to enhance entrepreneurial capabilities by supporting knowledge flows, particularly in South-South knowledge transfers and setting up financing schemes and arrangements. For example, China is increasingly becoming an important source of knowledge and is leading innovative activities in key areas (Wessner 2007). Bank Group financing represents a relatively small share of China's investment and financing needs, but it plays a prominent role in bringing ideas, knowledge, and best practice experience to help the country improve firm and sector competitiveness. The Bank Group's role as an honest broker is also valued. Thus, China has been requesting Bank Group support for projects in innovation and knowledge transfer.

In more recent CASs, upper-middle-income Brazil and China have focused on emphasized inclusiveness and requested Bank support for promoting inclusive innovation that addresses the needs of the poor. This is another area where there are promising opportunities for South-South exchange, because these countries share challenges and growth opportunities. Efforts to strengthen entrepreneurial capabilities have also included support to financing schemes such as venture capital funds and grants, as in the Mexico Innovation for Competitiveness Project.

Fostering linkages between the actors in the national innovation systems has been the least emphasized area in CASs that cover innovation and entrepreneurship. About a third of the CASs that IEG has reviewed addresses this issue to some extent.

Summary

Developing countries seeking to catch up and successful middle-income countries are seeking support from the World Bank Group to develop innovation strategies and policies that will strengthen their competitiveness, improve economic diversification, and stimulate growth. The Bank Group can help its clients build their innovation capacity so they can acquire, adapt, and use innovations that have been developed elsewhere. Experience from countries that have had successful innovation strategies may offer opportunities to help countries pursue innovation-driven growth. Bank Group CASs reflect increasing demand for innovation projects across different income categories.

However, current corporate and sector strategies do not provide adequate guidance on how to develop effective innovation interventions that can help client countries select, design, and implement policies and integrated programs to support innovation and entrepreneurship in a holistic manner. In fact, the World Bank Group does not have a comprehensive strategy or results framework for projects supporting innovation and entrepreneurship. Therefore, Bank Group interventions in this field have tended to be articulated around other thematic areas of interventions and not necessarily around innovation and entrepreneurship as a theme. This is partly because the agenda on innovation and entrepreneurship is still evolving.

Endnotes

[1] OECD conducted innovation policy reviews for selected nonmember countries—Chile (2007), China (2008), Korea (2009), Mexico (2009), and South Africa (2007); UNCTAD conducted 11 national Science Technology and Innovation Policy reviews as of the end of 2011—Angola (2008), Colombia (1999), Dominican Republic (2011), El Salvador (2011), Ethiopia (2002), Ghana (2010), Islamic Republic of Iran (2005), Jamaica (1999), Lesotho (2010), Mauritania (2009), and Peru (2010). In addition, as of 2011, there are requests for such reviews from seven countries—Ecuador, Iraq, Kenya, Pakistan, Papua New Guinea, the Philippines, and Sudan.

[2] Brazil, Chile, China, India, Kenya, Malaysia, Russia, Rwanda, Singapore, and Republic of Korea (Dahlman 2014).

[3] It is very challenging to find metrics that capture innovation, primarily because there is a dearth of official statistics measuring innovation—such as number of new products, processes, or other innovation. Most innovation measures also do not adequately capture the wide range of innovation outputs from the broad spectrum of innovation actors. These issues are discussed in INSEAD's *The Global Innovation Index* (INSEAD 2012).

[4] INSEAD's Innovation Input Sub-Index is built around five input pillars: institutions, human capital and research, infrastructure, market sophistication, and business sophistication. It captures elements of the national economy that enable innovative activities. Innovation outputs are the results of innovative activities within the economy. The Innovation Output Sub-Index has two output pillars—knowledge and technology outputs and creative outputs.

[5] Relative to the conceptual framework presented in Figure 1.1, these proxy measures of innovation (in Figures 2.1 and 2.2) are mostly at the level of immediate and intermediate outcomes, but related to the targeted interventions of the conceptual framework, in particular support to R&D (human capital and research) and strengthening entrepreneurial capabilities (offering training, business sophistication, quality certification).

[6] See *The Economist,* April 23, 2010, for an analysis of innovation in emerging markets.

[7] See Dahlman (2008) for discussion of innovation strategies in BRIC countries.

[8] ICT-enhanced innovations are spreading rapidly in developing countries, particularly in Asia and Africa. Examples of innovative applications of ICT include mobile banking in Kenya, the Philippines, and South Africa; applications in agriculture in Niger, Ghana, and India; and health services in India and other parts of Africa. Significant advances are expected as mobile phone technology and its application in innovative activities spreads, generating significant social and economic benefits.

[9] *The Economist* (2011) provides several examples of the growing importance of Diasporas and their contribution to a country's economic growth.

References

Arnold, E., and M. Bell. 2001. "Some New Ideas about Research for Development." In: *Partnership at the Leading Edge: A Danish Vision for Knowledge, Research and Development.* Copenhagen: Danish Ministry of Foreign Affairs.

Bell, M. 2007. "Technological Learning and the Development of Production and Innovative Capacities in the Industry and Infrastructure Sectors of Least Developed Countries: What Roles for ODA?" SPRU-Science and Technology Policy Research, University of Sussex, Paper prepared for UNCTAD Division for Africa, Least Developed Countries and Special Programmes.

Dahlman, Carl. 2014. "Innovation and Entrepreneurship: Framework, Lessons from International Experience, and Implications for the World Bank Group." Independent Evaluation Group Background Paper, Washington, DC.

——. 2008. "Innovation Strategies of the BRICS: Brazil, Russia, India, China, and Korea—Different Strategies, Different Results." PowerPoint Presentation, World Bank, Paris.

Devan, Janamitra. 2012. *The Innovation Imperative. Capital Finance International.* www.cfi.co/Africa/2012/12.

The Economist. 2011. "The Magic of Diasporas." November 19.

IEG (Independent Evaluation Group). 2012. *Adapting to Climate Change: Assessing World Bank Group Experience—Phase III of the World Bank Group and Climate Change.* Washington, DC: World Bank.

INSEAD. 2012. The Global Innovation Index: Stronger Innovation Linkages for Global Growth. Fountainbleau: INSEAD.

OECD and IDRC (International Development Research Centre). 2010. *Innovation and the Development Agenda.* Paris: OECD.

OECD and World Bank. 2009. *Innovation and Growth: Chasing a Moving Frontier.* Paris: OECD.

UNCTAD. 2011. *A Framework for Science, Technology, and Innovation Policy Reviews.* New York: UNCTAD.

Wessner, C. W., ed. 2007. *Innovation Policies for the 21st Century: Report of a Symposium.* Washington, DC: National Academies Press.

World Bank. 2012. *Agricultural Innovation Systems: An Investment Sourcebook.* Washington, DC: World Bank.

——. 2009. "World Bank Group Private Sector Development Strategy." Mid-Cycle Implementation Progress Paper, World Bank, Washington, DC.

Yusuf, Shahid. 2009. *Development Economics Through the Decades: A Critical Look at 30 Years of the World Development Report.* Washington DC: World Bank.

3

Supporting Innovation and Entrepreneurship in World Bank Group Projects

CHAPTER HIGHLIGHTS

- The World Bank Group has an investment portfolio of $18.7 billion related to innovation and entrepreneurship over the past decade, excluding non-lending activities.

- Bank Group projects on innovation and entrepreneurship are concentrated in lower- and upper-middle-income countries, but recently there is a shift to low-income countries.

- World Bank Group institutions designed and implemented four main types of interventions in response to identified market and government failures and other bottlenecks: support for R&D infrastructure, strengthening entrepreneurial capabilities, financing for early-stage start-ups, and fostering linkages among different actors.

- Both the Bank and IFC provided analytical and advisory support on innovation and entrepreneurship to clients.

The World Bank Group supports investment and advisory projects that help build innovation capacities and improve incentives for private enterprises to invest in innovations. These projects are intended to address the issues related to the enabling environment and the four targeted areas discussed in Chapter 1 (Figure 1.1). This chapter describes the characteristics of these projects and the types of interventions and mechanisms that have been used to implement them. The analysis is based largely on data from a portfolio review of projects that supported innovation and entrepreneurship based on IEG's criteria (Appendix B). The chapter also uses a country lens to get a better sense of the extent to which Bank Group activities fostered innovation and entrepreneurship at the country level. IEG examined the projects that were implemented during the evaluation period by the World Bank Group in five countries at different stages of development.[1] The choice of countries was based on the extent of Bank Group involvement in innovation and entrepreneurship projects, as well as the countries' inclination to undertake innovative initiatives.

In addition to its lending and non-lending portfolio, the World Bank supports innovation policies that complement its efforts through initiatives by the World Bank Institute and infoDev (Boxes 3.1 and 3.2). The World Bank Institute supports the Bank's operational work and its country clients with new approaches to capacity development. The Institute works in seven thematic areas: climate change, fragile and conflict-affected states, governance, growth and competitiveness, health systems, public-private partnerships, and urban development.

BOX 3.1 Knowledge and Innovation at the World Bank Institute

The World Bank Institute supports Bank clients in three main areas:

Open Knowledge: This area connects World Bank Institute clients to global knowledge and learning about the "how" of reform. In fields where content is mature, the Institute codifies global knowledge into training programs to help its clients test development know-how. Such courses can be found on the new e-learning platform, the e-Institute. The World Bank Institute also supports peer-to-peer learning and helps broker knowledge exchanges among developing countries. It encourages Bank country teams to incorporate knowledge exchanges in country programs and is promoting the Global Development Learning Network as a worldwide knowledge exchange implementation platform.

continued on page 43

Collaborative Governance: This helps clients mobilize for collective action by building multi-stakeholder coalitions that require effective and inclusive leadership as well as new forms of collaboration. The Institute offers four collaborative governance business lines:

• Open Government and Open Aid

• Capacity Building for Nongovernmental Actors

• Citizen Engagement through ICT

• Multi-stakeholder Collaborative Action.

Innovative Solutions: The Institute is developing tools, methods, and online platforms to facilitate an open and collaborative development process among governments, citizens, and other stakeholders. Its work in this area has three parts:

• *Open Data and Open Government:* The Institute has made data on more than 7,000 development indicators available for public use and in searchable, downloadable, and machine-readable formats. Examples of products that build on this are Mapping for Results and the Open Aid Partnership.

• *Competitions and Challenges:* The Global Apps for Development competition creates useful and innovative software applications using World Bank development data. Based on the competition, a new platform has also been customized that enables the World Bank to launch an array of competitions and challenges. The Institute also administers a $1.2 million Innovation Fund that supports World Bank staff in advancing ideas to improve development outcomes.

• *Scaling Social Enterprises:* The Bank launched the Development Marketplace in 2001 to position social entrepreneurs as the third arm of development, along with public and commercial private sectors. Since then, more than 300 global groups have won $200,000 each in grant funding. In 2011, the Development Marketplace was expanded with the launch of the Development Marketplace Investment Platform.

SOURCE: World Bank Institute.

The Information for Development program (infoDev) is a global partnership program and a key pillar of the ITE Global Practice in FPD. It pilots new initiatives on high-growth entrepreneurship, especially on business incubation and early-stage financing. Prior to its placement within the FPD Network, the program was a part of the Global Information and Communications department as a research, capacity-building, and advisory service focused on using ICT to help promote sustainable development and reduce poverty. infoDev collaborates with different parts of the World Bank Group that are working on entrepreneurship and is currently involved in four types of activities: business enablers, networks and capacity building, access to finance, and knowledge products (Box 3.2).

World Bank Group Lending and Investment Portfolio for Innovation and Entrepreneurship

WORLD BANK LENDING

Word Bank lending for projects that support innovation and entrepreneurship is directed toward governments in client countries. Funds are provided as direct support (loans and grants) to governments and are channeled to entrepreneurs through public sector institutions or public-private sector arrangements. In some cases, the private sector acts as the implementing agency for a government run project.

IEG identified a lending portfolio of 119 projects—64 closed and 55 active, located in 60 countries—that included activities relating to innovation and entrepreneurship with a total lending volume of $8.2 billion between FY00 and FY12.[2] Of these projects, 106 identified a lending amount that specifically supported activities related to innovation and entrepreneurship. Lending for closed Bank projects in this evaluation accounted for around 2 percent of total volume of Bank lending for projects exited during this period.[3]

Bank support for innovation and entrepreneurship projects responds to demand from its client countries. These projects, once concentrated in middle-income countries, are increasingly found in lower-income countries (Appendix Table D.2). If active projects are used as a proxy for recent lending efforts, the trend in distribution of lending activities suggests that Bank support for innovation and entrepreneurship may be shifting. The Africa Region had the largest number of projects, both closed and active. However, average lending per project in the region was the smallest (Appendix Table D.3). Among the sectors, FPD had the largest number of both closed and active projects (Appendix Table D.4). Figure 3.1 shows the distribution of projects in each Bank Group institution.

BOX 3.2 infoDev's Support for Innovation and Entrepreneurship

Business enablers (incubators, innovation centers, business acceleration programs): infoDev's incubation activities began with the Global Business Incubator Initiative in 2002. The goals of this initiative are to improve the performance of existing incubators and facilitate the development of new ones; promote knowledge generation and dissemination; foster national and international partnerships and networks; and foster ICT-enabled innovation. infoDev's incubation activities have since expanded to include agribusiness innovation centers, climate innovation centers, and mobile innovation programs (mLabs/mHubs, centers where entrepreneurs can find technical assistance, networking opportunities, and testing support for new applications). As of FY12, the network has nearly 240 incubators in more than 90 developing countries; these assist 20,000 enterprises. It also established four mLabs and has established climate innovation center programs in six countries.

Networks and capacity building: infoDev provides networking opportunities for entrepreneurs, private sector investors, and the donor community—through events (the annual Global Forum for entrepreneurs and SMEs), through business plan competitions and SME fairs, and through social networks. It also supports capacity-building initiatives that are targeted to policy makers, incubation managers, and trainers.

Access to Finance: Recently, infoDev launched a new program on Access to Finance. The program intends to design and pilot early-stage financing facilities. Some of the planned infoDev initiatives include Angel Co-investment and Technical Assistance, which consists of an early-stage innovation and financing facility; an incubator attached seed financing facility; and an innovative micro, small, and medium-size enterprise finance facility for the Caribbean.

Knowledge products: infoDev provides research to identify unique and innovative development opportunities and knowledge products, including policy guidance on approaches to licensing, competition, and universal access and on bandwidth sharing, mobile broadband, and net neutrality through good practice examples and benchmarks based on global experience. In FY10 and FY11, it commissioned or completed 19 policy-related studies (including five guideline manuals, five studies, three workshops, one focus group, and one case study).

SOURCE: infoDev.

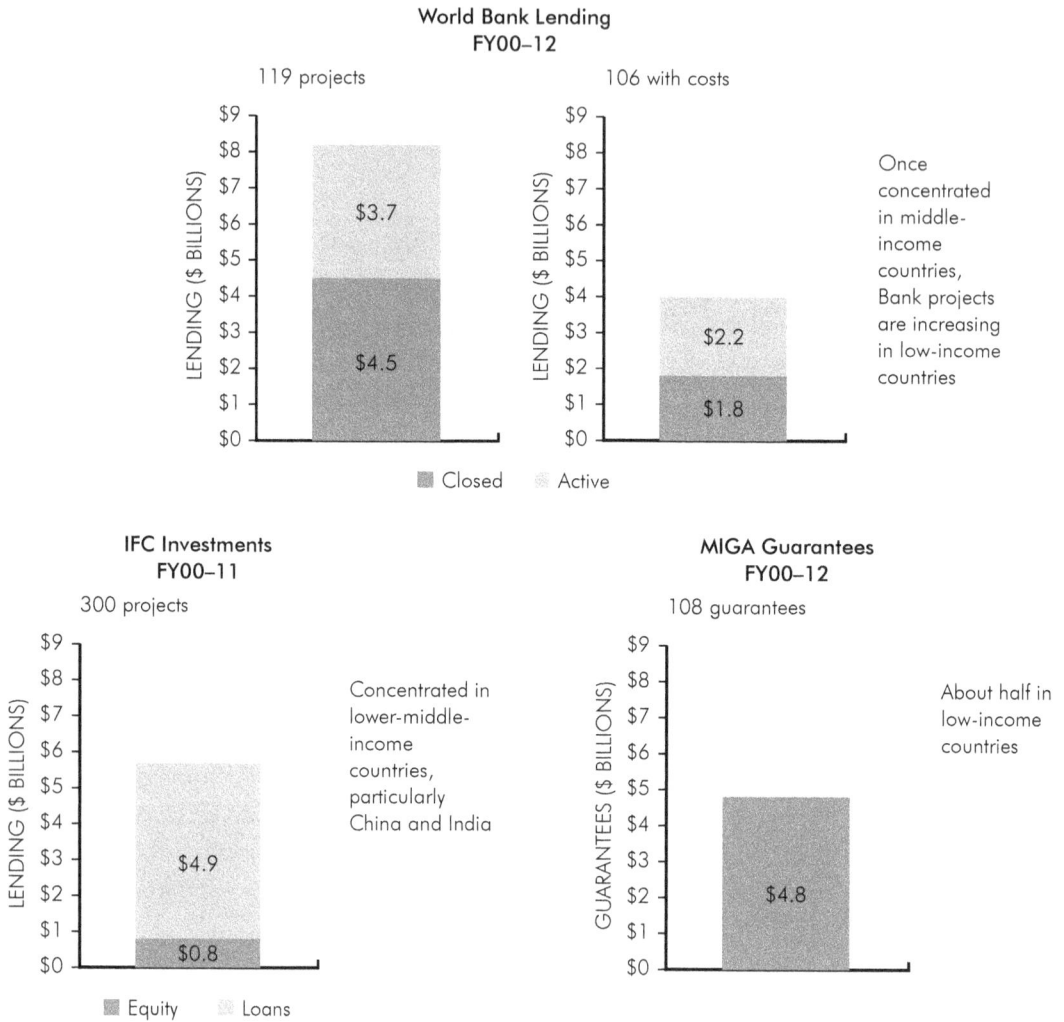

FIGURE 3.1 World Bank Group Lending Support for Innovation and Entrepreneurship

**World Bank Lending
FY00–12**

119 projects

106 with costs

$3.7

$4.5

$2.2

$1.8

LENDING ($ BILLIONS)

■ Closed ▧ Active

Once concentrated in middle-income countries, Bank projects are increasing in low-income countries

**IFC Investments
FY00–11**

300 projects

$4.9

$0.8

LENDING ($ BILLIONS)

■ Equity ▧ Loans

Concentrated in lower-middle-income countries, particularly China and India

**MIGA Guarantees
FY00–12**

108 guarantees

$4.8

GUARANTEES ($ BILLIONS)

About half in low-income countries

SOURCE: IEG.

IFC INVESTMENTS

IFC provides support for innovation and entrepreneurship by client companies, making investments directly in start-ups and existing companies that are willing to take risks and generate and/or disseminate new or improved products, processes, and marketing models. In this way, IFC's innovation and entrepreneurship investments are helping bring products, processes, services, and forms of business organization or marketing that are new to its client companies or even to countries in which its clients operate. In many investments, IFC also brings in foreign companies with considerable technological and business capabilities as co-investors or technical advisors, providing important channels and ideas for innovation.

IEG identified 300 IFC innovation and entrepreneurship projects in client companies located in 82 countries with total commitments of $5.7 billion between FY00 and FY11 (Appendix Table D.5).[4] Of the 300 investment projects considered, 203 had evaluation findings. These evaluated projects correspond to about 20 percent of the volume of all IFC projects evaluated during the study period.

Projects were concentrated in lower-middle-income countries, with about two-thirds of those in China and India (Appendix Table D.6). The number of these projects and investment commitment per project varied significantly across the regions. The Europe and Central Asia Region had the largest number of projects, followed by the Latin America and the Caribbean Region. East Asia and Pacific had the largest investment commitment per project, implying fewer projects but a larger volume of investment commitments (Appendix Table D.7). The top three sectors with these projects were manufacturing, financial markets, and agriculture and forestry (Appendix Table D.8).

MIGA GUARANTEES

MIGA provides political risk insurance to investors and lenders against noncommercial risks, primarily transfer restriction, expropriation, and war and civil disobedience. MIGA's guarantee coverage facilitates FDI that brings new products, processes, business organization, and innovations in marketing and distribution. These innovations are important because they have direct effects on businesses and consumers as well as significant demonstration effects when they are copied and replicated by local firms.

For this review, IEG identified 108 innovation-related projects in 53 countries, issuing $4.8 billion guarantees between FY00 and FY12 (Figure 3.1; Appendix Table D.9). The 108 MIGA projects accounted for about 30 percent of the number of guarantees issued during this period.

Forty-seven percent of investment guarantees issued were in low-income countries. Sub-Saharan Africa accounted for the largest volume, followed by Europe and Central Asia. These two regions accounted for about two-thirds of the total volume of guarantees. Innovation projects in the Latin America and the Caribbean Region had the largest average volume of guarantee issued per project. About 16 percent of projects involved South-South transactions. In terms of sector focus, the largest number of projects was in agribusiness. However, the largest volume of guarantees issued for innovation projects was in infrastructure, where projects tend to be large scale, with substantial investments in fixed assets.

Design of Bank Group Innovation and Entrepreneurship Projects

Productivity growth and competitiveness are important determinants of economic and social progress. Increases in productivity can arise from efficiency gains in existing businesses or reallocation of resources from less productive to more productive firms. However, market and government failures as well as other bottlenecks can pose multiple constraints to growth and development (Appendix A).

The World Bank, IFC, and MIGA design and implement different types of interventions that support entrepreneurship at different stages. The Bank helps governments address policy functions affecting innovation, such as providing support to innovative activity, reducing obstacles to innovation, funding relevant R&D, fostering dissemination and use, and supporting monitoring and evaluation (World Bank 2010). IFC and MIGA interventions focus mainly on private firms, assisting them with production, delivery, and scaling-up of new or improved processes, business models, and forms of marketing and business organization. About half of IFC's Advisory Services work is with governments. Each institution designs and implements interventions that are consistent with its mandate and comparative advantage, but the overall effort encompasses an array of interventions that can be used to foster innovation and entrepreneurship in a range of development contexts.

World Bank Group–targeted interventions to foster innovation and entrepreneurship consist of support to public and private R&D, strengthening entrepreneurial capabilities, financing early-stage start-ups, and fostering linkages between actors in the innovation system. Bank Group institutions emphasize different types of interventions, with the Bank focusing more on support for R&D infrastructure, strengthening entrepreneurial capabilities, and fostering linkages mainly between research universities and industry. IFC emphasizes firm-level issues such as strengthening entrepreneurial capabilities and financing early-stage start-ups, and MIGA focuses almost exclusively on strengthening entrepreneurial capabilities.

World Bank Interventions Supporting Innovation and Entrepreneurship

The World Bank has a diversified lending portfolio to support innovation and entrepreneurship. Targeted interventions have focused on building the R&D infrastructure and regulatory regime to develop new technologies and inventions; strengthening entrepreneurial capabilities; and, to a less extent, financing schemes and fostering linkages between public research systems and firms (Figure 3.2).

FIGURE 3.2 World Bank Interventions Supporting Innovation and Entrepreneurship

Number of projects
with intervention

Share
n = 119

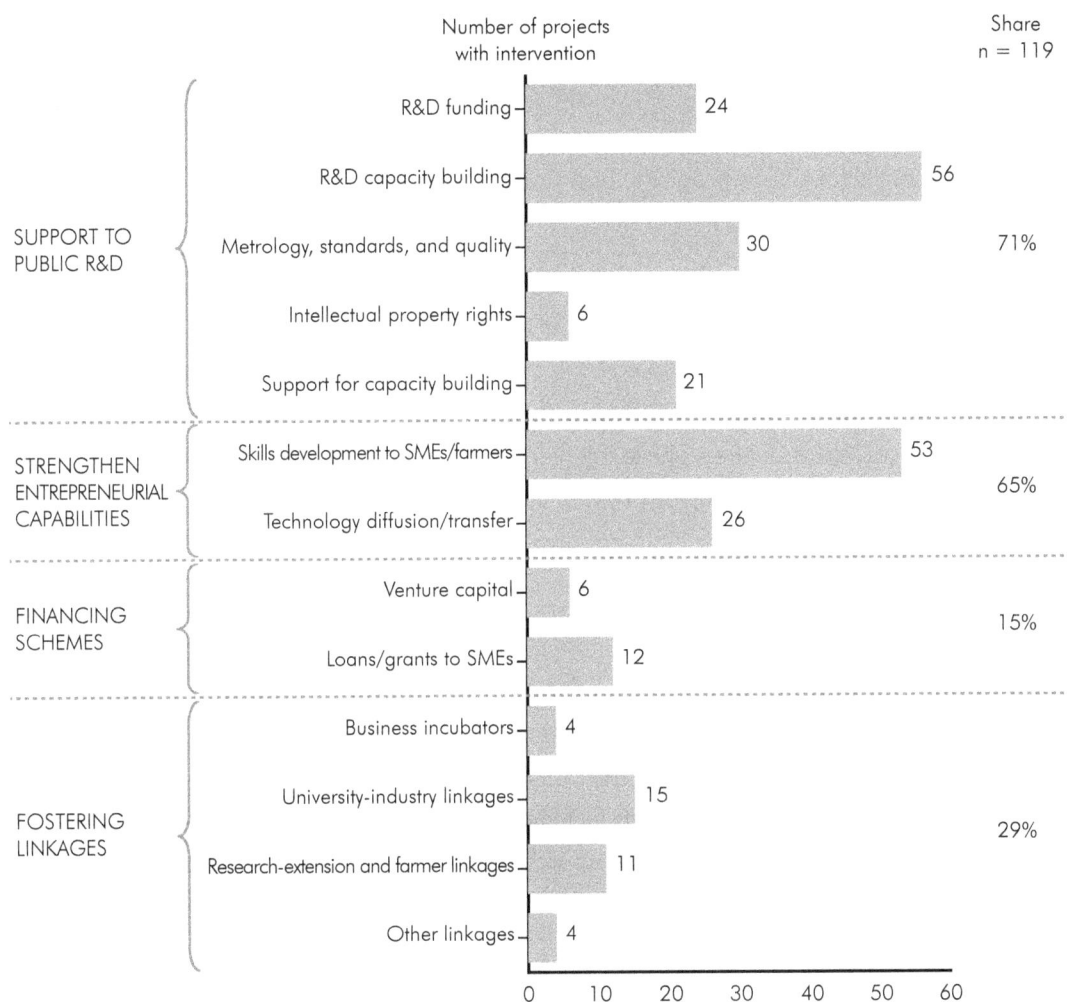

	Number of projects	Share
SUPPORT TO PUBLIC R&D		**71%**
R&D funding	24	
R&D capacity building	56	
Metrology, standards, and quality	30	
Intellectual property rights	6	
Support for capacity building	21	
STRENGTHEN ENTREPRENEURIAL CAPABILITIES		**65%**
Skills development to SMEs/farmers	53	
Technology diffusion/transfer	26	
FINANCING SCHEMES		**15%**
Venture capital	6	
Loans/grants to SMEs	12	
FOSTERING LINKAGES		**29%**
Business incubators	4	
University-industry linkages	15	
Research-extension and farmer linkages	11	
Other linkages	4	

SOURCE: IEG.

NOTE: R&D = research and development; SME = small and medium-size enterprise.

SUPPORT FOR PUBLIC R&D

The Bank provides support for public R&D because in the absence of countervailing institutions, particular types of market failures lead to the under-provision of certain types of public goods, such as research, that are necessary for the generation of basic knowledge and technologies that entrepreneurs can commercialize. To respond to inadequate provision of such public goods and services, the World Bank has provided funding to:

• Build or strengthen the technological infrastructure for R&D in public research institutions and S&T parks to conduct basic and applied research to develop new and improved technologies

- Support public universities, particularly science and mathematics departments and research labs

- Build or strengthen national quality infrastructure, including institutions, laws, and regulations

- Support capacity for policy, program, and strategy related to innovation capability and competitiveness.

Support for research and technological infrastructure has been used to develop and diffuse new and improved technologies that help improve firm-level productivity and strengthen competitiveness (Figure 3.2). Eighty-five of the 119 innovation projects supported by the Bank—71 percent of all projects—included activities that supported investments in research and technological infrastructure.[5] These interventions were concentrated in low-income countries, mainly through funding for physical infrastructure, research facilities, and capacity building of research and scientific staff. Priority was given to capacity building of research staff aimed at promoting S&T outputs. In middle-income countries, there was more emphasis on strengthening national quality infrastructure.

Bank-supported R&D was mostly in ARD, education, and FPD. In agriculture, for example, the Peru Agricultural Research and Extension Project (1999) sought to increase the productivity and competitiveness of the agricultural sector through the adoption of environmentally sound technologies. Project components included an agricultural technology fund that provided competitive research grants to research and extension institutions and institutional strengthening activities to help build capacity in the national technology system.

Some education interventions focused on building scientific capacity to enhance innovation. For example, the Chile Millennium Science Initiative's (1999) objective was to improve the performance of the Chilean S&T system. Project components included a competitive fund for scientific excellence and the creation of a network to promote scientific excellence. In addition to the competitive funds, other mechanisms used to implement these interventions included technical assistance for selected science institutes; funding for scientific infrastructure, equipment, and fellowships; and exchange programs with advanced research institutions.

In FPD, the focus was mainly on strengthening national quality infrastructure in public institutions. For example, the objective of the Ghana Private Sector Development Project (1994) was to foster the development of a competitive private sector by helping the government develop appropriate technology and improved knowledge of quality and standards. The project enabled capacity building to strengthen the role of Ghana National Standards Body in developing national quality infrastructure and disseminating these services across the country. Another project, the Nicaragua Micro, Small, and Medium-Size Enterprise Development Project (2008), supported quality and certification services by strengthening the national quality control laboratory and Ministry of Health certification office.

The incentive problem is another issue that impedes innovation; some innovative individuals and firms do not attempt to gain the full benefits of an innovation because others can easily copy their ideas. Firms in a competitive environment may under-invest or not invest at all in transforming R&D outputs into commercial products when they cannot prevent other investors from imitating and providing their innovation without getting any financial benefits. The World Bank addressed incentive problems by supporting IPR regimes to better connect firms to products from R&D and R&D funding.

Six projects, accounting for five percent of innovation projects, incorporated interventions that addressed IPR regimes, mainly in upper-middle-income countries. IPR regime interventions are designed to provide the incentive to commercialize inventions. These interventions helped promote IPR regimes in client countries to bring them in line with international standards. The expectation is that clearer and better-enforced IPR would provide incentives for entrepreneurs to commercialize R&D products, resulting in more innovations being brought to market. In the Turkey Industrial Technology Project (1999), the Bank helped strengthen industrial property rights and services by supporting improved patent filing and search examination procedures; establishing dedicated information centers for documentation and information dissemination; and enforcing industrial property rights.

STRENGTHENING ENTREPRENEURIAL CAPABILITIES

Firms play a central role in the innovation process. In some cases, information asymmetries result in firms and potential entrepreneurs not having the necessary information to seize potential business opportunities. Beyond market failures, capability failures can be another key bottleneck in innovation processes. Managerial deficits, lack of technological understanding, and limited learning ability or "absorptive capacity" to make use of externally generated technology capabilities may exist. To address these failures, Bank Group interventions provided support to help build managerial and entrepreneurial capabilities in firms and support for enterprise innovation and upgrading, including introduction of new products into markets, technology transfer, and technology diffusion.

Sixty-five percent of innovation and entrepreneurship interventions supported strengthening of entrepreneurial capabilities, with a majority of these supporting skills development for SMEs and farmers. Technical assistance and capacity building have focused on management training, helping firms acquire skills through experience, and business development services and providing a wide range of support to help firms start or run a company. These interventions were concentrated in low- and lower-middle-income countries, where they accounted for more than two-thirds of all Bank support for strengthening entrepreneurial capabilities.

An important focus of World Bank Group interventions was on strengthening entrepreneurial capabilities in SMEs, with a view to improving firm-level competiveness, growth, and access to markets. For example, the overall objective of the Uganda Private Sector Competitiveness Project (2004) was to create sustainable conditions for enterprise creation and growth that responded to local and export markets. Project components included enhancing enterprise competitiveness aimed to improve enterprise capacity by encouraging investment in skills; raising productivity; and improving the quality, standards, and reliability of micro, small, and medium-size enterprise producers in export value chains. Mechanisms used to implement this intervention included a matching grant scheme and competitive grant scheme for business plan development. In Nicaragua, the Micro, Small, and Medium-Size Enterprise Development Project's (2008) objective was to improve the competitiveness of such firms and the business climate affecting them. Matching grants were used to help introduce new products or processes and reduce the time needed to start a business. These grants also financed other activities, such as quality enhancements and certification, innovation, labor training, and clean technologies.

Some Bank interventions have supported the introduction of new climate change products into markets. For example, climate change innovations are being introduced in many markets. In a World Bank project in Sri Lanka, output-based aid and innovative financing arrangements were used to successfully introduce a new renewable energy technology—solar photovoltaic home systems—in the market (IEG 2010). Several water efficiency projects in China used satellite-based measures of crop evapotranspiration, an innovative approach, to measure actual water use (IEG 2012).

FINANCING SCHEMES

A problem with innovation investments is that they are often risky, with uncertain outcomes. Such investments are not attractive to banks and other financing institutions, so they are less likely to provide financing. In addition, many innovative projects have a relatively low probability of success and may turn in profits over a longer time frame than conventional financing institutions expect. This inherent uncertainty of success results in limited financing; consequently, firms may under-invest in innovative projects.

In response, the Bank Group has provided financial support for early-stage start-ups through venture capital funds, as well as loans and grants to innovative and entrepreneurial companies and SMEs. However, this has not been a major area of focus for the Bank. For example, IEG identified six venture capital funds—four in upper-middle-income and one each in low- and lower-middle-income countries—that the Bank supported in the innovation and entrepreneurship lending portfolio. One such fund was a component in the Argentina Unleashing Productive Innovation Project (2008); it promoted the development of new knowledge-based companies by establishing a pilot venture capital fund for technology sectors with an emphasis on early-stage financing. The objective of the venture capital fund was to make early-stage finance available for technology start-ups and to provide a demonstration effect to show that these investments were commercially viable. The Bank also supported entrepreneurs through grants and soft loans for concept development, business planning, export support, and company accreditation.

FOSTERING LINKAGES BETWEEN THE ACTORS IN THE INNOVATION SYSTEM

Technological development and innovation processes are complex, so effective linkage between different policies and relevant actors in the innovation system is critical for success. Linkages can be among domestic and foreign firms and with universities, research institutions, and technology intermediaries, as well as with the Diaspora. Bank Group activities supporting linkages included the following:

- Research-university industry linkages

- Research-farmer-research extension linkages

- Business incubators, providing a range of services to start-ups and young firms and linking them

- Platforms such as the Innovation Policy Platform developed by the World Bank and OECD.

Thirty-four projects—29 percent of World Bank projects—fostered linkages between the actors in the innovation system. Around half of these linkages were between industry and the private sector. Their objective was to commercialize R&D products that have been produced in public R&D institutions and universities. Most of the interventions fostering linkages were in agriculture, but some others sought to strengthen linkages between academia and industry. For example, the objective of the India National Agricultural Innovation Project (2006) was to contribute to the sustainable transformation of the Indian agricultural sector from a primary focus on food self-sufficiency to one based on market orientation. One of the project components involved establishing research consortia to facilitate dialogue and interactions among public research organizations, farmers, private sector, and other stakeholders to support agricultural transformation. Mechanisms to implement project interventions included competitive grants for research activities, grants for establishing research consortia, and capacity building.

In education, the Uganda Millennium Science Initiative (2006) supported universities and research institutes to produce more and better-qualified science and engineering graduates and higher-quality and relevant research, and enhanced linkages between research and industry. One project component provided grants for "technology platforms" through which private firms and researchers defined collaborative agendas and pursued relevant solutions to issues faced by industry.

The commercialization of R&D outputs involves the transformation of inventions into new products, processes, or services that start-ups or existing companies can develop and bring to the market. The Croatia S&T Project (2005) is a good example. One of the project programs supported collaboration between the private sector and research institutes through a matching grants scheme that financed 20 projects. A new generation of projects has put much more emphasis on promoting S&T that is linked to competitiveness and industry productivity. Such projects include the Chile Science for the Knowledge Economy Project (2003), the Mexico Innovation for Competitiveness Project (2005), the Uruguay Innovation Loan (2007), and the Argentina Unleashing Productivity Innovation Project (2009).

The World Bank has also supported business incubators by offering them a variety of support resources and services. Business incubators link innovation and entrepreneurship and help bring new ideas to the market, contributing to jobs and economic growth. When successful, they can create strong linkages among financiers, universities, policy makers, and firms (Khalil and Olafsen 2010). The Bank has provided funding for government operation or subsidized business incubators to help start-ups and innovative SMEs commercialize their innovation and grow into successful firms. For example, the Bank supported the establishment of the Enterprise Incubator Foundation in Armenia in 2002 to help develop its information technology sector. The project provided a comprehensive package of services to Armenian information technology firms through business linkage services, skill development services, and managed workspace. More recently, the Bank and IFC have supported business incubators through grants made to infoDev (Appendix Box D.1).

The Bank Group's ICT department has facilitated collaboration among governments, development partners, and the private sector to leverage external sources of knowledge and expertise. For example, the Bank helped structure a partnership between Moldova and Singapore in which Singapore provided technical assistance to help design Moldova's e-government efforts. A knowledge platform on ICT was developed as a joint initiative between the Bank's ICT unit and the World Bank Institute to focus on linking clients and staff with external sources of knowledge and expertise in the sector (IEG 2011).

The World Bank and OECD jointly developed the Innovation Policy Platform to foster the use of innovation policies and programs to increase sector and firm competitiveness across industries and countries (Box 3.3). It is a Web-based open data interactive platform aimed at facilitating collective learning processes around STI policies. Its goal is to provide its users—innovation policy makers and practitioners globally—with support in analyzing innovation systems and policies and in shaping future policy design.

BOX 3.3 World Bank–OECD Innovation Policy Platform

The Innovation Policy Platform will mobilize global resources, knowledge, and expertise to help policy practitioners learn about various elements of innovation policy design, implementation, and monitoring and evaluation, as well as identification and prioritization of the good practice solutions most appropriate for their contexts. This open-data interactive platform will facilitate knowledge exchange and peer-to-peer learning among policy makers and practitioners in developed, emerging, and developing countries through:

- An open data portal of up-to-date knowledge on innovation policy globally

- Interactive search networks and communities of practice to locate explicit and tacit knowledge and skills and identify solutions to specific innovation policy needs

- Feedback and peer review structures to enhance the learning opportunities and allow the platform to be an active instrument for policy debates.

The project that will build this platform involves a better codification and packaging of existing innovation knowledge.

SOURCES: OECD and World Bank 2009; World Bank FPD.

Bank projects used different mechanisms to implement interventions that support innovation and entrepreneurship. These mechanisms included competitive fund mechanisms, such as competitive research grants and matching funds. In many cases, several mechanisms were combined to implement an intervention.

Support for Public R&D

Competitive Research Grants. The competitive research grant (CRG) is an important mechanism that has been used to help improve performance and efficiency in public research systems, improve the research-industry link, and promote private sector participation in public sector research. With CRGs, research providers are selected on a competitive basis, based on technical proposals and peer review.

At the World Bank, CRGs have been used mainly to support agricultural innovation, although they have also been used to improve the quality and relevance of higher education and to strengthen linkages among national quality infrastructure, standards bodies, and private industry in PSD. Eighteen Bank projects used CRGs to improve performance in public research systems; the ARD and education sectors used it most frequently. In ARD, CRGs were often linked to agricultural research funds that supported agricultural research, technology transfer, and extension, as well as the provision of agricultural services. Education projects have used CRGs to improve the quality and relevance of education in public R&D institutions and universities.

Training and Technical Assistance. This kind of assistance activity has supported capacity building in S&T for basic and applied research and national quality infrastructure in public research institutes and enhanced public research and/or university linkages with industry. Main mechanisms include scholarships and grants for training at master's, doctoral, and postdoctoral levels and twinning arrangements and other forms of collaboration with international research institutions, laboratories, and universities.

Strengthening Entrepreneurial Capabilities and Linkages

Matching Grants. Entrepreneurs play an important role in commercializing R&D products that have been developed through public R&D institutions and universities. The World Bank has provided subsidies to help firms commercialize R&D products developed in public research institutions. Matching grants, in which the Bank provides a partial subsidy to firms, have been used to facilitate development of new products through collaboration between firms and R&D institutions. Nine projects have used this mechanism to facilitate such collaboration, with the expectation that grant funding would provide incentives for entrepreneurs to bring innovations to market.

Matching grants have also been used to help entrepreneurs finance the cost of business development services, export promotion activities, and technology upgrading. This mechanism was used in 23 projects to support business development and consulting services, mainly by FPD.

The Bank Group also supported innovation and entrepreneurship through the World Bank Institute's Development Market Place.

IFC Investments Supporting Innovation and Entrepreneurship

IFC's innovation and entrepreneurship projects focus almost exclusively at the firm level, with interventions that aim to strengthen entrepreneurial capabilities mostly through incentives for firm-level growth through technological upgrading and financial support for early-stage start-ups. IFC's investments have supported firm expansion and growth through technology upgrading. This occurs through four main channels: technology transfer or technology diffusion, upgrading existing products and processes, firm-level R&D for product development, and introduction of innovations into the market. Seventy-four percent of projects have supported technology upgrading efforts mainly by helping firms upgrade existing products and processes and technology exchange (see Figure 3.3).

IFC's financing for acquiring new technology and for technology transfer interventions are mainly supported by its manufacturing sector, accounting for 45 percent and 35 percent of such interventions, respectively. Agribusiness and forestry also support about a quarter of all interventions that help firms acquire new technology and technology transfer. Through these interventions, IFC helps link companies to a global pool of technology, knowledge, human capital, and learning by doing.

IFC interventions supporting firm-level upgrading and modernization efforts as well as introduction of new products, processes, and institutions were concentrated in lower-middle-income countries. For example, IFC provided long-term foreign currency debt financing to a leading Indian pharmaceutical company (2009) to support its expansion into China. The project involved South-South technology exchange and knowledge transfer. In another project, IFC supported a leading soft drink bottler and distributor to transfer a successful business model from Asia and Africa. This investment also enabled the company to transfer its technology from one frontier market to another.

FIGURE 3.3 IFC Interventions to Support Innovation and Entrepreneurship

Number of projects
with intervention

Share
(n = 300)

	Number of projects with intervention	Share (n = 300)
STRENGTHEN ENTREPRENEURIAL CAPABILITIES	Technology transfer/diffusion — 92	
	Upgrading existing products and processes — 108	
	R&D for product development — 22	74%
Introduction of innovation to market	Establishment of new institutions — 34	
	Introduction of new products and services — 25	20%
FINANCING SCHEMES FOR EARLY-STAGE START-UPS	Early-stage financing through venture capital fund — 12	
	Early-stage financing directly to the company — 7	6%

0   20   40   60   80   100   120

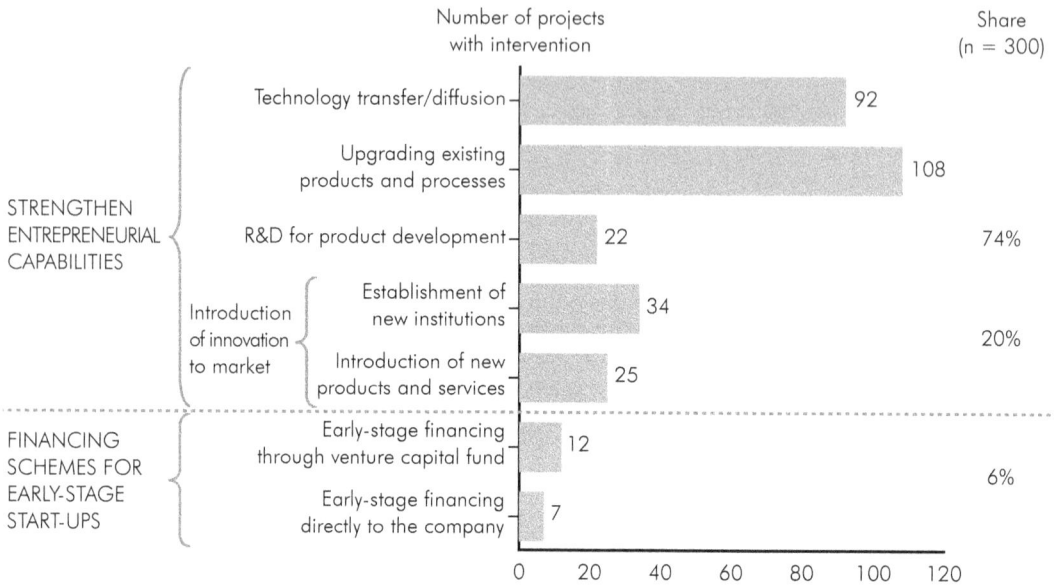

SOURCE: IEG.
NOTE: R&D = research and development.

Innovations also occur when new products, processes, or marketing or organizational models are introduced. IFC has provided loans and equity support for start-up companies and innovative firms that are willing to take risks to introduce new products and services. Twenty percent of IFC's interventions helped firms introduce innovation into markets, mainly through the establishment of new financial institutions and the introduction of new products or services. These interventions were mainly in financial markets: 94 percent of cases where financial institutions were established and 68 percent of cases where client firms introduced new products, such as equipment leasing and credit bureaus into markets.

IFC investments have fostered innovation by helping start-ups and innovative firms introduce new products, services, and business models to the market. For example, IFC has supported clients that introduced new leasing operations in Peru and Tanzania, new insurance products in the Middle East and North Africa, software technology in new export markets, new products and flexible pricing schemes in the information technology sector in Paraguay, new financial products in many countries in all regions, and clean energy technology or energy efficient products in several regions.

Both the World Bank and IFC have supported index insurance schemes that offer the advantages of insurance to farmers and livestock keepers at lower cost than traditional approaches (IEG 2012). Some of these interventions involve new ways of delivering financial services to underserved segments of the population. In many cases the innovations are intrinsically inclusive. For example, IFC's support for a micro finance institution in the Democratic Republic of the Congo reduced the cost of opening a bank account, increased access to banking facilities, and enhanced the affordability of a wide range of financial services to previously underserved populations.

The returns to investments by innovative firms may be high, but capital markets may not provide long-term capital for risky ventures with uncertain outcomes. This is particularly difficult for new firms, because they do not have a track record or collateral, which banks require for making loans. The problem is compounded when the firms are start-ups based on new, untried technology that they have developed or are trying to implement for the first time. Success may bring high financial gains or negative returns. For such cases, IFC has invested in venture capital funds that pool and manage money from investors who take private equity stakes and invest in start-up companies and SMEs with strong growth potential.

IEG identified a subset of 12 venture capital funds from IFC's broader equity portfolio that focused on early-stage companies and innovative SMEs mainly in lower-middle and upper-middle-income countries.[6] Many of these funds were small, ranging from $2.5 million to $25 million. In some cases, IFC has taken a place on the board of directors of these funds. The focus on risk financing for early-stage start-ups is important because start-ups with untried technologies or without a market track record tend to experience greater access constraints to finance, posing acute barriers to their growth (Dahlman 2014).

IFC investment in venture capital funds provides early-stage companies and innovative SMEs with equity capital as well as managerial expertise, market information, and other forms of technical assistance. For example, IFC invested in a private equity fund that targeted early-stage venture equity and quasi-equity investment opportunities in Indian high-technology and high-growth equity companies. The project was created to help small and medium-sized companies by providing scarce capital, managerial talent, and market information. IFC has also invested in regional funds that involved establishing a venture capital fund to make equity and quasi-equity investments in private sector small and medium-sized companies in the South Pacific Island countries. In a few cases, IFC has also provided equity and quasi-equity investments directly to start-ups and high-growth SMEs.

MIGA Guarantees Supporting Innovation and Entrepreneurship

MIGA's efforts to promote FDI in developing countries can play a vital role in fostering innovation and entrepreneurship. By providing coverage for political risk insurance, its interventions directly address incentive problems that may cause firms to under-invest in innovative products and processes. The main channels through which MIGA's support for FDI fosters technology upgrading in client firms were technology transfer (in 37 percent of innovative projects) and acquisition of new production technology and processes (in 28 percent of innovative projects). Technology-upgrading interventions provided important channels for the flow of technologies and know-how between a foreign investor and client in a developing country.

MIGA's support for innovative interventions focuses on strengthening entrepreneurial capabilities by facilitating firm growth and expansion through (i) transfer of technology or equipment, (ii) transfer of business process or practice, and (iii) capacity building through training or knowledge transfer. The bulk of the technology upgrading interventions were in infrastructure, with the sector accounting for 50 percent of support for technology transfer and 40 percent of acquisition of new technology and processes. For example, MIGA issued a guarantee in a project that supported the construction and operation of a seawater desalination plant in China. The project involved the transfer of technology and know-how on advance water treatment technology from Norway to China through a joint venture enterprise.

In another project, MIGA supported South-South investment by providing a guarantee contract for a project to design, construct, and operate the first geothermal plant in Kenya. Through its support for foreign private investment in this project, MIGA helped introduce geothermal technology, know-how, and managerial expertise to Kenya. In another South-South transaction, MIGA's insurance coverage helped investors from India acquire new and simple technology for roofing products. In all these cases, MIGA's insurance coverage was critical in facilitating technology upgrading, innovation, and knowledge flows through technology transfer, technology diffusion, and acquisition of new technology that supported firm growth and expansion.

In 35 percent of projects, MIGA supported client firms in introducing new products and processes into the market. These interventions included support for the establishment of new financial institutions, such as the first leasing company in Serbia and Montenegro, or support for a new mobile banking services and payment system in Sierra Leone. Financial services dominated interventions that introduced innovations into markets, accounting for about two-thirds of such interventions.

World Bank Group Knowledge Activities Supporting Innovation and Entrepreneurship

WORLD BANK ADVISORY AND ANALYTIC ACTIVITIES

The World Bank supported client countries through AAA comprising economic and sector work (ESW) and technical assistance. A review of a random sample of 250 closed and active AAA projects implemented between FY00 and FY11 found that 36 percent of these projects involved innovation and entrepreneurship projects. Of these, ESW dominated, accounting for 66 percent of such work.

These studies focused on broad issues such as innovation policy, knowledge economy, and technology, mainly to inform government policies. Technical assistance, accounting for 34 percent of innovation-related AAA, focused on strengthening institution and client capacity to implement innovation projects. The Sustainable Development Network accounted for 53 percent of technical assistance projects, and the Poverty Reduction and Economic Management Network had the most ESW projects (39 percent) (Appendix Figure D.1).

In a portfolio of 18 active AAA projects, 13 involved technical assistance, 4 ESW, and 1 both lending and ESW. On average, $393,222 was spent on the four ESW and $328,731 on the technical assistance projects. These projects show an increasing share of FPD in AAA projects, with the sector accounting for 8 of the 18 studies. Nine of 18 AAA projects were delivered to clients in middle-income countries, whereas only 2 of all the active AAA in innovation-related studies were in low-income countries. ESW outputs were mainly knowledge reports and policy notes intended to inform government policy or stimulate debate on various aspects of innovation policy. Technical assistance in innovation-related activities mainly involved diagnostic work, providing assistance in strategy implementation, policy guidance, institutional capacity building, and raising awareness to facilitate knowledge exchange.

IFC ADVISORY SERVICES

IEG identified 84 IFC Advisory Services projects between FY05 and FY12—58 closed and 26 active—that had innovation-related interventions that supported entrepreneurship.[7] At the design stage, IFC Advisory Services projects are expected to identify the market failure or firm-level constraint that the project is addressing.

Total expenditure on the 84 projects that supported entrepreneurship was about $42 million, with most spent in middle-income countries. Slightly more than half of this expenditure, $23 million, was attributed to the Access to Finance business line, and about $15 million was spent by the Sustainable Business Advisory business line (Appendix Table D.10).

Advisory Services innovation projects supported innovation and entrepreneurship through three major types of interventions: building entrepreneurial capabilities, management training and skill development, and institutional building or policy reform. Of these interventions, support for building capabilities in start-up and innovative SMEs was most frequent. Access to Finance and Sustainable Business Advisory business lines accounted for the majority of interventions that helped build entrepreneurial capabilities in innovative firms.

Support for building entrepreneurial capabilities was implemented with mechanisms such as technical assistance to help firms with feasibility studies, product development, and growth strategies. Innovative SMEs, such as those involved with energy-efficient technologies, received support to help them develop and bring energy-efficient products to market. In other cases advisory services were provided to support capacity building in innovative firms. Through these grants, IFC Advisory Services supported commercialization of products from R&D, such as clean energy technologies, helping entrepreneurs bring innovations to markets.

Country Perspectives: Design of Innovation and Entrepreneurship Interventions

IEG analyzed intervention at the country level for all the projects implemented by the World Bank, IFC, and MIGA in Brazil, Chile, China, India, and Kenya over the evaluation period (Figure 3.4). Together these countries accounted for 27 percent of the number of Bank projects, 23 percent of IFC projects, and 10 percent of MIGA projects reviewed in this evaluation.

Brazil, Chile, and China are upper-middle-income countries and are considered leaders in pursuing innovation-driven growth in their national development strategies and CASs. The Bank Group institutions also have a track record of consistently supporting innovation and entrepreneurship projects in these countries.

India and Kenya are in the lower-middle and low-income categories, which only recently have begun to give innovation priority in national strategies and CASs. Yet both are making important strides in innovative activities, particularly in incremental and inclusive innovations that provide solutions to pressing development challenges. All these countries were included in the 10 country case studies in Chapter 2, thus providing important linkages with key principles that can be used to foster innovation and entrepreneurship.

FIGURE 3.4 Innovation and Entrepreneurship Interventions in Selected Countries

Country	Institution	Innovation Type			
		Public R&D Support	Strengthening Entrepreneurial Capabilities	Financing Schemes	Fostering Linkages
Brazil	IFC (n = 13 projects)		▣	▪	
	MIGA (n = 5 projects)		▪		
	WB (n = 5 projects)	▪	▪		▪
Chile	IFC (n = 3 projects)		▪		
	WB (n = 3 projects)	▪	▪		▪
China	IFC (n = 16 projects)		▣		
	MIGA (n = 4 projects)		▪		
	WB (n = 4 projects)	▪	▪	▪	
India	IFC (n = 32 projects)		■	▪	
	WB (n = 3 projects)	▪	▪		▪
Kenya	MIGA (n = 4 projects)		▪		
	WB (n = 2 projects)		▪		▪

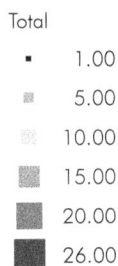

Total

- ▪ 1.00
- ▣ 5.00
- ▣ 10.00
- ▣ 15.00
- ■ 20.00
- ■ 26.00

SOURCE: IEG.

NOTE: Figure includes only World Bank lending, IFC investments, and MIGA guarantees.
Sum of total (size) broken down by innovation type versus country and institution. The view is filtered on innovation type, which excludes null.

The CASs developed over the evaluation period for countries such as Brazil, Chile, China, and India included innovation and entrepreneurship as pillars or strategic priorities for achieving broader development objectives like strengthening competitiveness and growth.

In Chile, investments in research and innovation have been important priorities to address the country's global competitiveness since the 1990s; these priorities continue to be important pillars in the current CAS. Country strategies in Brazil and China included innovation as a key strategic objective for strengthening competitiveness and expanding services to the poor. Innovation and knowledge, seen as crucial for delivering Brazil's growth agenda, have been central elements in the Bank's policy dialogue with the government. China's CAS intends to accelerate the pace of innovation by creating an open innovation system in which competitive pressures encourage Chinese firms to engage in product and process innovation through their own R&D as well as through participation in global R&D networks.

In India, inclusive, sustainable growth and service delivery were strategic priorities in earlier CASs, with innovation emphasized in the agricultural sector, mainly to shift the focus to commercially oriented agriculture and public-private partnerships. More recently, innovation has taken a more central role in addressing the country's development challenges, guided by government initiatives that focus on projects that transform and modernize policies and institutions, leverage resources, and pilot new and innovative development approaches.

Kenya's early CASs mention innovation and entrepreneurship but mainly in agriculture, with the goal of achieving an innovative, commercially oriented, competitive, and modern agricultural sector. Recent CASs, however, acknowledge the importance of research in informing government policy debates and the importance of the Bank Group in introducing innovative solutions to the country's development problems.

COUNTRY ANALYSIS BY TARGETED INNOVATION AND ENTREPRENEURSHIP INTERVENTIONS

Figure 3.4 shows the World Bank, IFC, and MIGA's targeted interventions supporting innovation and entrepreneurship in the five countries selected for this analysis.

Support for R&D

The Bank has provided consistent support for R&D in all these countries, but there are important differences in the design and content of R&D interventions. In Brazil, the emphasis has been on supporting public R&D infrastructure (public research institutions; metrology, standards, and quality control infrastructure; and regulation including IPR). Chile, in contrast, emphasizes pilot initiatives to build scientific excellence and scale up the successful components into more integrated support for human capital development (master's, doctoral, and postdoctoral work), support for policy and strategy, and stronger linkages between universities and industry.

Another model used in China supported R&D by promoting the development, adaptation, and commercialization of new technologies and standards through institutional development, strengthening national quality infrastructure, and offering study tours. These efforts included a focus on enhancing the promotion of innovation related to the environment and energy efficiency. In India, support for R&D has focused mainly on the agricultural sector and innovation systems perspectives, with researchers, farmers, and other stakeholders getting involved in setting the agricultural research agenda and public-private partnerships playing a key role in implementing sector priorities.

Bank projects supporting R&D interventions are concentrated in education, ARD, and FPD. But there are also sector differences in the content of these interventions. For example, education projects in Brazil emphasize human capital development, whereas in Chile they emphasize high-quality scientific capacity and more recently broader innovation system perspectives, including scientific capacity and research-industry linkages. ARD projects in Brazil, India, and Kenya have supported the development of R&D infrastructure and human capital in agricultural S&T development. In India, projects have also embraced innovation system perspectives that emphasize linkages between research and other actors in the innovation system. FPD projects have supported components of R&D that focus on incentives to commercialize products from R&D by strengthening IPR regimes and national quality infrastructure.

The World Bank also supported analytical work to inform strategies and investments in innovation and entrepreneurship in Brazil, Chile, and China. At the federal level in Brazil, a study was undertaken on knowledge, innovation, and competitiveness. Two studies on innovation systems addressed firm and regional competitiveness in China. Three projects in Chile aimed to strengthen government policies on innovation and inform a national strategy for innovation.

Strengthening Entrepreneurial Capabilities

World Bank projects in these five countries have provided support for skills development, mainly to enhance the capacity of entrepreneurs and SMEs to interact with technology development and/or adopt standards that could increase market access. In Brazil, Bank projects supported capacity building to enterprises in a project on S&T reform support and technology adaptation and diffusion. In China, World Bank efforts to strengthen entrepreneurial capabilities focused on helping firms tap into global knowledge and technology through technology transfer. In three of the four World Bank projects, there was a focus on accelerating the pace of innovation by helping Chinese firms participate in global R&D networks.

The bulk of interventions that sought to strengthen entrepreneurial capabilities were supported by IFC and MIGA, mainly through technology upgrading, knowledge flows, and skills enhancement via technology transfer, technology diffusion, and acquisition of new technologies and processes. In Brazil and Chile, IFC investments helped with the expansion and modernization of firms' production facilities, services, and distribution networks and technology transfer, aiming to increase productivity, reduce cost, and improve efficiency. MIGA interventions that helped strengthen entrepreneurial capabilities in Brazil helped its clients establish and upgrade power networks and upgrade production facilities in agribusiness, manufacturing, and service sectors. In China, IFC's interventions helped upgrade existing products and processes and introduce new products and services, technology transfer, and firm-level R&D for product development in manufacturing, consumer and social services, and telecom and ICT.

In India, IFC investments, in some cases combined with Advisory Services, have played important roles in introducing new renewable and green products, such as solar power plants, solar roof tops, and energy-efficient street lightning. MIGA's guarantees supported its Chinese clients in upgrading existing products and processes and technology transfer in the water, waste water, and transportation sectors. In Kenya, MIGA also supported clients that introduced new products and helped firms upgrade their technology and processes, facilitating technology and knowledge flows as well as enhancing innovation.

Financing Schemes

The five-country experience shows that a range of financing mechanisms has been used to support innovation and entrepreneurship. However, there appears to be institutional specialization; each institution uses some mechanisms more frequently. The Bank has used grants and loans, whereas IFC has used loans, equity, and a combination of these. MIGA exclusively uses guarantees to support its clients.

The World Bank has used CRGs almost exclusively to support R&D in Brazil, Chile, China, and India. Matching grants were used to reduce risks and provide incentives to entrepreneurs in Bank projects in Chile, China, and India. Both the Bank and IFC have supported venture capital schemes. But IFC has been more active in supporting venture capital initiatives, particularly in India. IFC has also provided advisory services that helped financial institutions expand their focus on clean energy financing.

Fostering Linkages between Innovation Actors

The World Bank fostered linkages between research and industry, mainly in FPD and education projects in Brazil and Chile. In Brazil, projects supported partnerships among industries, universities, technological institutes, and government agencies and helped establish university-business innovation networks. In Chile, mechanisms such as research consortia were used to strengthen research-industry linkages. However, the limited number of such interventions in the five countries' innovation and entrepreneurship investment portfolio suggests that the Bank Group needs to do more to help countries focus on transferring scientific results and technologies to develop solutions that address development challenges in specific contexts.

ARTICULATING INNOVATION INTERVENTIONS AT THE COUNTRY LEVEL

World Bank Group interventions from projects reviewed in each of the five countries are summarized in Figure 3.4. Analyses of these interventions on a country basis show that:

- **Brazil:** Within the Bank Group, the World Bank was the only institution supporting R&D in Brazil. All Bank Group institutions supported entrepreneurial capabilities, but with different emphases. The Bank emphasized skills development, whereas IFC and MIGA helped strengthen firm capabilities through diverse technology upgrading efforts. The Bank also supported interventions to enhance linkages between universities and the private sector. Targeted interventions for financing entrepreneurs were not a major part of Bank Group support for innovation and entrepreneurship.

- **Chile:** The World Bank supported two projects in R&D, the first piloting interventions in building scientific excellence and the second emphasizing scale-up of relevant S&T activities, innovation strategy, and linkages between universities and industry. The Bank and IFC supported entrepreneurial capabilities but, like in Brazil, with different emphasis on skills development and technology upgrading. The Bank Group did not provide any targeted financing for entrepreneurs in Chile.

- **China:** The World Bank supported R&D in environment projects, including renewable energy and energy efficiency, consistent with the government priorities on environmental sustainability and technology transfer. All the World Bank Group institutions supported entrepreneurial capabilities, mainly by helping firms tap into global knowledge and technologies. Targeted financing for entrepreneurs was not a major part of Bank Group support for innovation and entrepreneurship.

- **India:** World Bank support for R&D focused mainly on the agricultural sector and innovation systems perspectives, as reflected in the CAS priorities. Other interventions sought to strengthen linkages among researchers, farmers, and other stakeholders in the agricultural innovation system. IFC was active in strengthening entrepreneurial capabilities and financing schemes for entrepreneurs.

- **Kenya:** Both the Bank and MIGA supported entrepreneurial capabilities, with a similar pattern in which the Bank emphasized skills development and MIGA technology upgrading. There were no interventions supporting R&D and targeted support for entrepreneurs.

The analysis of project experiences from a country perspective reveals a number of important insights on Bank Group support for innovation and entrepreneurship at the country level. First, the way innovation is addressed in CASs tends to influence investment in innovative activities. Countries that identify innovation and entrepreneurship as pillars or strategic priorities in their CAS tend to emphasize such projects in investment projects from different World Bank Group institutions.

Second, Bank Group interventions are broad and cover key elements in building innovation capacity at the country level. These interventions tend to be designed and implemented as stand-alone activities by specific sectors and different Bank institutions. Key areas, such as targeted financing for entrepreneurs and fostering linkages between actors in the innovation system, do not appear to get adequate consideration in country interventions.

Third, given the systemic nature of innovation processes, the extent to which Bank Group interventions can achieve their full potential at the country level depends on whether there are mechanisms that facilitate coordination, knowledge sharing, and joint action across different sectors and Bank group institutions.

Summary

The World Bank Group has a significant and diversified portfolio of activities to foster innovation and entrepreneurship. Current support is concentrated in lower- and upper-middle-income countries. Yet there is growing recognition that innovation is important at all stages of development. World Bank innovation-related interventions have traditionally focused on building an environment conducive to business development, supporting the R&D infrastructure, and supporting skills development and training in firms. IFC and MIGA have provided support at the firm level, mainly through technology transfer, upgrading existing technologies, and introducing new products and process. Knowledge created through analytical work also plays an important role in strengthening Bank Group policy advice on innovation strategies and policies. Analysis of Bank Group interventions from country perspectives shows that innovation and entrepreneurship interventions are designed and implemented by sectors and Bank Group institutions in ways that may not be necessarily connected.

Endnotes

[1] The countries are Brazil, Chile, China, India, and Kenya.

[2] The $8.2 billion total lending includes all project components, some of which may not relate to innovation and entrepreneurship.

[3] A reviewer noted that World Bank annual lending for innovation and entrepreneurship projects corresponds to less than 0.3 percent of the Bank's annual lending for Development Policy Operations.

[4] The investment portfolio comprised mature projects that were evaluated between FY00 and FY11, as well as recent projects approved between FY08 and FY11.

[5] World Bank projects typically have more than one intervention or activity contributing to the achievement of their main or component objectives. Hence the number of interventions is greater than the number of projects.

[6] IFC supports entrepreneurs by filling gaps in equity financing through a broad portfolio of private equity funds, including SME funds, midcap, venture capital, mining, health care, clean tech, and agribusiness. Through these private equity funds, IFC facilitates asset building, capital formation, and growth in new firms and innovative SMEs at different stages in the commercialization cycle.

[7] These projects were either approved or evaluated between FY05 and FY12.

References

Dahlman, Carl. 2014. "Innovation and Entrepreneurship: Framework, Lessons from International Experience, and Implications for the World Bank Group." Independent Evaluation Group Background Paper, Washington, DC.

IEG (Independent Evaluation Group). 2012. *Adapting to Climate Change: Assessing World Bank Group Experience— Phase III of the World Bank Group and Climate Change.* Washington, DC: World Bank.

———. 2011. *Assessing IFC's Poverty Focus and Results.* Washington, DC: World Bank.

———. 2010. *Climate Change Phase II: The Challenge of Low-Carbon Development.* Washington, DC: World Bank.

Khalil, Mohsen A., and Ellen Olafsen. 2010. "Enabling Innovative Entrepreneurship through Business Incubation." In Lopez Claros (ed.) *The Innovation for Development Report 2009–2010.* New York: Palgrave Macmillan.

OECD and World Bank. 2009. *Innovation and Growth: Chasing a Moving Frontier.* Paris: OECD.

World Bank. 2010. *Innovation Policy: A Guide for Developing Countries.* Washington, DC: World Bank.

4

Portfolio Performance of World Bank Group Support for Innovation and Entrepreneurship

CHAPTER HIGHLIGHTS

- Most World Bank projects designed to support innovation and entrepreneurship achieved their stated outcomes. These projects have mainly supported R&D interventions, which have traditionally been a mainstay of the Bank's innovation agenda.

- Country development context is associated with differences in the performance of Bank but not IFC projects.

- In IFC, investment projects that supported innovation and entrepreneurship had significantly lower development impact and returns than other projects evaluated during the same period. This contrast is associated with differences in work quality and higher levels of risk in interventions.

- The performance of MIGA's innovation and entrepreneurship projects was no different than of other MIGA projects, but they had lower ratings for quality of assessment, underwriting, and monitoring.

- Several mechanisms have been used to implement innovation and entrepreneurship interventions across the Bank Group. These have been most effective when used in areas with a track record, such as competitive research grants to improve efficiency in public research systems or support for technology upgrading in firms.

- Performance has been lower in high-risk, high-reward areas such as financing early-stage start-ups through venture capital funds or fostering linkages among innovation actors. There is mixed experience on the use of matching grants to improve firm performance.

The World Bank Group institutions have designed and implemented diverse interventions, ranging from support to building an enabling environment to targeted interventions related to R&D infrastructure, to improving the linkages between various innovation actors, to helping build innovation capacity and foster entrepreneurship in client countries. Evaluation findings indicate varying degrees of success across interventions. This chapter assesses the overall and specific performance of the targeted interventions indicated in Chapter 1 that Bank Group institutions have used to support innovation and entrepreneurship. The assessment of performance is based on projects for which IEG has evaluative evidence: 65 of the 119 World Bank projects, 203 of the 300 IFC projects, and 18 of the 108 MIGA guarantees in the portfolios identified in Chapter 3. This evaluative evidence is derived from Implementation Completion and Results Reports, Expanded Project Supervision Reports (XPSRs), and Project Evaluation Reports (PERs) generated by the project-level evaluation process for each of the three World Bank Group institutions. These data sources were supplemented with Project Performance Assessment Reports and Evaluative Notes, which together ensures representativeness in the case of IFC and the Bank, but not for MIGA, whose project-level evaluation system is young and evolving.

IEG assessed performance at the aggregate project level by examining the extent to which innovation and entrepreneurship projects achieved their intended development objectives. This analysis also includes an assessment of specific mechanisms that were used to implement the interventions. IEG conducted case studies to provide further insights on the performance of different interventions by assessing their benefits and challenges from the perspectives of start-ups, innovative existing firms, and key stakeholders. This chapter presents an analysis of achievements in CASs for the five countries examined in Chapter 3. This analysis is patchy because several CASs did not provide a comprehensive analysis of the achievements of different Bank Group interventions.

Performance of World Bank Projects

ASSESSMENT OF PROJECT OUTCOMES

Project outcome ratings, linked to a project's objectives, provide an idea of the extent to which a project's major relevant objectives were achieved, or were expected to be achieved, efficiently.[1] This assessment is anchored to IEG's ratings of project outcomes, providing one perspective on the success or failure of projects to attain their intended project outcomes.

IEG reviewed 64 closed World Bank projects that supported innovation and entrepreneurship. Most Bank projects had objectives to increase sector and firm-level competitiveness directly or indirectly through innovation or technology development. Of the evaluated projects,

56 percent had a major or subobjective to directly increase competitiveness. In addition, 31 percent of projects intended to promote innovation or improve technology.

About 80 percent of the completed projects had satisfactory outcomes, achieving the major relevant objectives, most of which related to improving innovation, productivity, or strengthening competitiveness. Project performance was slightly higher than other Bank projects evaluated during this period (77 percent success rate), but this was not statistically different (Appendix Figure E.1).[2] Innovation and entrepreneurship projects were less successful in low-income countries than in lower-middle and upper-middle-income countries. Such differences suggest that the performance of interventions depends on the local context.

Bank projects typically involve several components, some of which may not relate directly to innovation and entrepreneurship. Thus, aggregate measures of performance, such as project outcome ratings, may reflect the effects of other interventions. IEG therefore divided project support between major and minor innovation and entrepreneurship projects.[3] Of the 64 projects, 35 were considered major innovation and entrepreneurship projects and 29 minor projects.

Analysis of major and minor innovation and entrepreneurship projects suggests that the intensity of effort on innovation and entrepreneurship activities was associated with achievement project outcomes (Appendix Table E.1). Major innovation and entrepreneurship projects were significantly more successful in achieving their major objectives than minor ones (89 percent versus 69 percent). Furthermore, the difference in successful achievement of project objectives between major and minor projects holds across country income levels (Appendix Table E.2). One explanation for the difference is that most of the major projects tend to be in agriculture and education, sectors where the Bank has had a stream of investments and developed a strong track record with successful performance.

IEG disaggregated efficacy ratings by objective to get a better sense of the extent to which objectives directly related to innovation and entrepreneurship were achieved. The most common objective was improving competitiveness, but projects also included other objectives, such as improving the enabling environment and enhancing access to finance. Of the 200 objectives referring to each outcome in the project development objectives, 117 (59 percent) were directly related to innovation and entrepreneurship. Within sectors, 93 percent of all education, 78 percent of agriculture, and 43 percent of FPD projects had relevant objectives that supported innovation and entrepreneurship. This also suggests that Bank effort supporting innovation and entrepreneurship has been dominated by the education and agriculture sectors (Appendix Table E.3).

Innovation and entrepreneurship-related objectives were more likely to be achieved than other project objectives (60 percent versus 54 percent), even though this difference was not statistically significant. Further analysis of efficacy ratings suggests that differences in achievement of objectives are due mainly to diverse sectoral performance. For example, both education and ARD projects were successful in achieving innovation and entrepreneurship objectives in more than 60 percent of cases, whereas the corresponding rating for FPD was slightly more than 50 percent (Appendix Table E.4). This difference in sector performance is also related to the strong track record that the Bank has developed in supporting S&T projects, which dominated earlier innovation projects.

ASSESSMENT OF BANK AND BORROWER PERFORMANCES

Bank performance, in terms of the quality of project design and supervision, and borrower performance in preparing and implementing projects are key determinants of project outcomes. Overall, innovation and entrepreneurship projects had lower performance ratings than other Bank projects that were evaluated during the same period. However, these differences were not statistically significant, suggesting that the quality of project design and supervision by the Bank and government and implementing agency performance in client countries is no worse than other Bank projects (Appendix Figure E.3).

Differences in country contexts, captured by country income categories, were associated with Bank and borrower performance. Across the Bank Group, in low-income countries, Bank and borrower performance in innovation-related projects was much weaker than other projects evaluated during the same period. However, in upper-middle-income countries, Bank and borrower performance in innovation-related projects was much stronger than in other projects in this income category (Appendix Figures E.4 and E.5). Thus, country contextual similarities—manifested in levels of economic development, technological capabilities, and institutional capacity—were associated with differences in performance. Bank and borrower performances were strongly correlated with achievement of successful outcomes. About 82 percent of innovation and entrepreneurship projects with satisfactory borrower performance achieved satisfactory outcomes, and 90 percent of such projects with satisfactory Bank performance achieved satisfactory outcomes. Given the importance of Bank and borrower performance in determining project objectives, IEG further explored the attributes of project-level variables related to performance (Appendix Tables E.8A and E.8B).

Innovation and entrepreneurship projects that did not achieve their objectives were more likely to be associated with poor project design than projects that successfully achieved their objectives. Key factors associated with lower performance in these projects were inadequate supervision; inadequate risk assessment (risk factors not identified at project design); overly complex designs, such as inclusion of multiple objectives; inadequate technical design, such as failure to identify clear links between inputs and outcomes; and inadequate borrower performance. Projects with unsatisfactory performance were just as likely to have inadequate risk assessment and weak monitoring and evaluation (M&E) as were those with successful outcomes (Table 4.1).

The main problems with project performance were associated with the Bank's role, irrespective of whether projects achieved their objectives. The issues were with project design (complex design, unrealistic targets, inadequate M&E) and quality of supervision. On the

TABLE 4.1 Factors Associated with Project Performance in IFC Investment Projects

| Performance Issue | Projects with Unsatisfactory Outcomes | | Projects with Satisfactory Outcomes | | Ratio[a] |
	Number	%	Number	%	
Inadequate supervision	8	62	5	10	6:1
Overly complex design	6	46	14	27	3:1
Lack of stakeholder involvement	1	8	2	4	2:1
Inadequate technical design	10	77	20	39	2:1
Inadequate risk assessment	3	23	3	6	4:1
Inadequate M&E framework, poor data quality/indicators	10	77	31	61	1:1
Inadequate skill mix of Bank team	3	23	0		
Inadequate borrower performance	11	85	9	18	4:1
Implementation disrupted by a crisis	4	31	8	16	2:1
Number of projects	13		51		

SOURCE: IEG.
NOTE: M&E = monitoring and evaluation.
[a] This is a ratio of percent unsatisfactory to percent satisfactory outcome.

borrower side, problems were caused by inadequate performance of government and implementing agencies and implementation delays.

A number of interesting features emerged from this analysis. On design, inadequate technical design appears almost as often in successful projects as in unsuccessful ones. As many projects with inadequate M&E fail as those that succeed. On implementation, problems occurred on both the Bank and borrower side. Also, all projects were affected by implementation problems. Setbacks occurred not only in projects that did not achieve their development outcomes but also in projects that successfully achieved them.

IEG used multivariate analysis of the outcomes of innovation and entrepreneurship projects to determine whether some constraints were more binding than others. IEG found that borrower performance is a key determinant of the achievement of a project's objective. Even good supervision cannot adequately compensate for poor borrower performance. The quality of Bank supervision is important, but project supervision and design have complementary effects. A project that is not well designed is less likely to achieve its objective even with good supervision. In contrast, a good design is not enough to ensure the achievement of project objectives when the project is poorly supervised. Appropriate targeting and effective M&E also work together. The achievement of a project's objective is threatened in the absence of one or the other (Appendix F). The interaction term results are robust to the nonlinearity of the model, as highlighted in Ai and Norton (2003).

Performance of World Bank–Targeted Interventions

Aggregate measures of project performance discussed above may include activities that do not directly support innovation and entrepreneurship. Of the 64 projects in this review, nearly half were classified as minor innovation and entrepreneurship projects. Major innovation and entrepreneurship projects also embedded activities that may not be innovation related. IEG disaggregated project components to capture relevant project activities, classified by the main types of interventions that the Bank Group has used to support innovation and entrepreneurship.[4]

The analysis of component data provides additional evidence of the bias toward S&T in Bank support for innovation and entrepreneurship. Component ratings by intervention indicate that the majority of project components relating to innovation and entrepreneurship provided support to public R&D, with 77 percent successfully achieving component objectives (Table 4.2). Project components fostering linkages had the highest successful ratings, but there were relatively few of these activities. As noted, the distribution of effort and performance of various innovation and entrepreneurship interventions reflects an early emphasis on correcting market and government failure in innovation projects. This perception of innovation

TABLE 4.2 Component Performance of Innovation and Entrepreneurship Projects

Intervention	Number of Components	Successful (%)
Support to public R&D	83	77
Strengthening entrepreneurial capabilities	41	73
Financing schemes	4	50
Fostering linkages in innovation system	10	80

SOURCE: IEG.

justified Bank support for R&D and building capability of firms to facilitate the transfer and commercialization technologies developed from R&D or acquired from foreign sources.

PERFORMANCE OF SPECIFIC MECHANISMS USED IN WORLD BANK INTERVENTIONS

IEG assessed the performance of the main mechanisms that have been used to implement the four types of innovation and entrepreneurship interventions.

Support to R&D

CRGs have been used to improve performance and relevance in public research institutes and universities. This mechanism was effective in improving transparency and efficiency of funding and delivering outputs, mainly in ARD and education interventions. Seventy-four percent of the projects using CRGs had satisfactory or better ratings for delivery of outputs. In ARD, 13 of the 15 projects successfully used the mechanism to promote high-quality research and technology transfer, encourage the private sector to participate in delivering agricultural services, and encourage users to participate in priority setting, funding, and delivery of services. All the education projects with CRGs used the mechanism effectively to either promote high-quality scientific research in public research institutions and/or universities or strengthen collaboration among research institutions, universities, and industry. CRGs were most effective in interventions in which there were transparent and rigorous selection procedures and strong institutional capacity in the research system, and where additional investments were made in capacity building activities to improve proposals from weaker institutions.

Training and technical assistance activities have been used to help build capacity of research and university staff in S&T. ARD interventions have used scholarships and grants for long-term training at master's, doctoral, and postdoctoral levels in agricultural S&T. In other cases, training has been effective in boosting the capacity of weaker institutions to participate in CRG programs.

Training and technical activities have been less effective in introducing reforms in public sector agricultural research institutes. In such cases, the lack of a strategic plan for reform, challenges posed by competing interests—public research institutes, universities, or line ministries—in the reform process, and staff resistance contribute to limited success in achieving reform objectives.

FPD has successfully used mechanisms such as staff training to help public research institutes develop commercialization strategies that respond to the needs of the private sector and to strengthen the role of standards bodies. The use of twinning arrangements with advanced metrology and calibration laboratories was effective in setting and disseminating measurement standards, and staff training has strengthened the capacity of standard bodies, increasing their efficiency and helping them provide enhanced support to industry and trade.

Strengthening Entrepreneurial Capabilities

Matching grants were used in 21 projects to support interventions that sought to strengthen entrepreneurial capabilities and provide funds for commercialization of R&D products. Fifteen of these projects were implemented by FPD; other sectors accounted for the rest. Thirteen of the 21 projects successfully delivered project outputs. In FPD, 9 of the 15 projects that used matching grants delivered outputs that provided access to business development or consulting services, improved know-how and knowledge that improved firm productivity, and increased access to export markets. The mechanism was effective under the following circumstances: the selection process was rigorous and followed clear procedures, project design was flexible to deal with changing circumstances, processing of claims for reimbursement was expeditious, and the private sector or private sector associations were involved in the administration of the scheme.

In 8 of the 21 projects, matching grants were not effective; that is, they did not deliver the expected outputs or achieve the intended outcomes. Project evaluations suggest that the failure of the grants was associated with several factors:

• A failure to correctly identify target beneficiaries

• Slow and costly implementation (for example, in one project 8 percent of the funds were disbursed but 100 percent of management fees was spent)

- Low uptake (for example, in another project, 45 percent of funds were disbursed by the end of the project)

- Problems with eligibility criteria that were either too strict and excluded firms that could benefit or too lax so participating firms were not selected well

- Complex processing and reimbursement procedures that caused cash flow problems in firms

- Rigid donor budgetary and procurement processes that caused delays in start-up

- Political interference that may have led to frequent changes in management

- Unfavorable macroeconomic conditions that created uncertainties for private sector investments.

These design and implementation challenges have meant that even when firms do get funds from matching grants, a large number of them do not proceed with the activities that were proposed.

Some interventions combine different mechanisms to implement innovation and entrepreneurship interventions. Box 4.1 provides an example of matching grants combined with competitive research grant to develop and commercialize new technologies that are having impact in those societies.

BOX 4.1 Harnessing Innovation for Improving Access to Electricity in Rural China

Geographically isolated rural populations in western regions of China faced a severe lack of access to electricity in the late 1990s, which inhibited their ability to reduce poverty and improve life. Solar home systems and small off-grid solar stations (collectively, PV systems) offered a potentially promising solution. But the industry was nascent, with low quality and high prices. This affected affordability and hampered demand.

continued on page 80

The 2001 China Renewable Energy Development Project, with support of about $40 million from a World Bank loan and a Global Environment Facility grant, used cost-sharing or matching grants to support technology improvement activities in manufacturers that supply PV components and to support efforts in improving efficiency, quality, and market development by assemblers in nine provinces. The project also provided a subsidy for PV system sales to the provinces, on the condition that the product and component quality met project standards, thus giving further incentives for manufacturers and assemblers to improve technology and quality. The project helped establish PV standards and testing and certification centers. Because all support was based on a competitive selection process, the project enhanced competition among companies in the industry. This reinforced the incentives for innovation and entrepreneurial activities.

The project contributed markedly to technology and product quality improvements. At project completion, 197 technology improvement activities were carried out, and 95 percent met or exceeded their contract targets. Seventy-four component manufacturers met project quality standards; some of these companies later became major players in the international market. The 18 most active assemblers developed the capacity to offer PV systems that meet international standards, and most of them became ISO-9001 certified. The project introduced higher technical standards and strengthened China's capacity in testing and certification for PV components and systems. The capabilities of four accredited PV testing centers had been strengthened to international standards by 2006.

Field interviews confirmed that the project stimulated successful innovation and technology adoption. For example, manufacturers invested the grants from the project and their own counterpart funds to adopt and develop new technologies to improve product reliability and efficiency, as well as to develop new products. Assemblers noted that the project helped them conduct innovative market development activities, enhance quality control, and renovate assembly lines.

In addition, the average price of PV systems declined during the project implementation period, after adjusting for inflation. The project supported competition and cost reduction, which contributed to price decline. Assemblers also reported that competition helped drive the PV system price down. In fact, more than 400,000 PV systems were sold to isolated rural areas in the provinces under the project.

SOURCES: Field survey and interview; IEG 2010.

Public sector support can help development of a venture capital industry in developing countries. In Mexico, a pilot venture capital scheme was established as part of a project in 2006. The fund was successfully established and provided resources for five start-up companies involved in information technologies, auto parts manufacturing, and infrastructure technologies. Although an S&T project in Croatia originally envisaged development of a local capital venture capital fund, this component was dropped in restructuring of the project. External factors such as the global economic and financial crisis played a big role in the cancellation of the venture capital funds. The fund also had difficulties identifying a portfolio of prospective investments.

Fostering Linkages among Innovation Actors

Matching grants have also been used to encourage collaborative links among research institutions, universities, and firms, with a view to facilitate the introduction of new products and processes into markets. For example, in Croatia a matching grant was effectively used to strengthen collaboration between research and industry to help commercialize products from R&D (Box 4.2).

BOX 4.2 Research Commercialization—Collaboration between the Public and Private Sectors

The economy of Croatia in 2005 had limited R&D commercialization and infrequent cooperation between enterprises and research institutions. The government decided to develop a new and comprehensive S&T policy (in line with its accession to the European Union) and asked for World Bank support.

The project the Bank supported had two objectives: to enhance public financing for business R&D and to foster the commercialization of public R&D. The Sponsored Research and Development Program (SPREAD), one of the programs in the project, aimed to support collaboration between the private sector and research institutes through a matching grants scheme that financed 20 projects. The majority of companies financed by the scheme engaged in computer programming, consultancy, and related activities. Companies that applied for SPREAD financing were small and micro R&D-oriented companies. Following project closure, the Institute of Economics in Zagreb assessed the effectiveness of SPREAD, using surveys and interviews with beneficiary and potential beneficiary SMEs in technology-intensive sectors.

continued on page 82

The study found that most of the SPREAD-financed projects resulted in new products or services (89 percent); they also improved existing processes (33 percent) and products (28 percent) and developed new processes (22 percent). Seventy-two percent of companies cited affordable financing of R&D projects as one of the benefits from the SPREAD program, as well as an improved competitive position nationally (61 percent) and internationally (56 percent). SPREAD recipients perceived that the scope of the R&D activity would have been smaller without the program, and the duration of the project would have been considerably longer without SPREAD support. However, companies indicated a low level of acquaintance with the program, indicating the need for improvement in the communication strategy for SPREAD.

A majority of beneficiary and potential beneficiary companies had no prior experience in collaboration with research institutions. They had more experience collaborating with other companies on their R&D projects. Two main motives for beneficiary companies' participation in the SPREAD scheme were an opportunity to alleviate financial risks related to R&D activities and an opportunity to increase each company's competitive advantage.

The SMEs strongly agreed that the SPREAD program facilitated their R&D activities. The program was also considered a facilitator of collaboration between SMEs and scientific institutions. One reason for the success of the project was the Business Innovation Centre of Croatia, whose role in supporting and developing ideas into financeable projects was critical for the development of projects.

SOURCE: Project evaluation documents.

In the Knowledge and Innovation Project in Mexico, the Bank supported technology transfer units at universities and National Council for Science and Technology centers. Technical assistance and investment in information technology was used to promote outreach to the private sector with a view to strengthening linkages between universities and industry. By the end of the program, about one-quarter of the technology transfer units supported had filed a total of 24 patents and a similar proportion demonstrated that they could serve as a bridge for collaborative research with industry.

Incubators are another mechanism that has been used to support the entrepreneurial process with a view to increase survival rates for start-ups and innovative SMEs. They offer entrepreneurs physical space, management coaching, and help in developing effective business plans, administrative services, technical support, business networking, legal advice, and advice on financing (www.infoDev.org). The World Bank has provided direct support for business incubators, for example, in Armenia. More recently the Bank and IFC have supported incubators through grants to infoDev. The basic idea addresses the commercialization gap of getting innovative ideas from public R&D labs and universities to start-ups and existing innovative firms. Available evidence suggests that the success of business incubators is mixed. For example, a review of the general performance of incubator programs in the Europe and Central Asia Region indicates that they have not been effective in terms of successfully advancing businesses or cost-effectiveness (Goldberg and others 2011). It is not clear, however, the extent to which the findings of this review can be generalized beyond Europe and Central Asia because of different contexts, incentives, and practices.

Evidence on the effectiveness of specific World Bank Group–supported incubators is limited because there have been relatively little rigorous evaluations that assessed the performance of firms that exit incubators, compared to those that did not use its services. An analysis of lessons from infoDev's support for business incubators in developing countries suggests that the initiative has positive effects on project outputs and outcomes (infoDev 2006).[5] Evidence from the study suggested that infoDev's business incubation program played an important role in helping grantees to succeed, which in turn helped their client firms generate economic and social benefits (Box 4.3). However, there is not much that can be said about the impact of infoDev's incubation interventions on the basis of this study because it did not specify useful comparisons and benchmarks.

BOX 4.3 infoDev's Support for Business Incubators

infoDev's study on effectiveness of its incubation program involved 49 grantees who responded to a survey. Of all respondents, 47.8 percent strongly agreed and 37 percent agreed with the statement that without the infoDev grant their organization could not undertake the proposed activities that the grant funded.

Even though just under half of grantees reported that in general access to ICT infrastructure is a challenge in their business environments, 88.4 percent reported that more than 76 percent of their staff had access to the Internet, and almost all reported some type of Internet connection; 93 percent of grantees indicated they have a website. Grantees indicated that they provided a variety of services to their clients using the Internet, including research and reference material (14.5 percent), access to email and the Internet (13.1 percent), publishing business opportunities (13.1 percent), and provision of toolkits (11.7 percent). Almost half of grantees reported that more than 75 percent of the total new jobs their clients created were ICT enabled.

SOURCE: infoDev.

IEG's own case study, based on the perspectives of firms in a World Bank–supported incubator in Armenia, points toward the mixed performance that has been observed elsewhere (Box 4.4). These insights demonstrate some successes but also major challenges that limited the performance and growth prospects of firms in the incubator.

Performance of IFC Investment Projects

The development performance of IFC investment projects was assessed on four measures: development outcome, investment outcome, project business success, and PSD impact.[6] Both project business success and PSD impact feed into the development outcome rating. Among projects evaluated during the same period, the cohort of interventions that supported firm-level innovation had significantly lower development and investment outcome success ratings than other projects (Appendix Table E.9). IEG did not identify any significant differences in IFC projects' success by country income level. A larger percentage of projects in lower-middle-income countries had mostly successful or better ratings than those in low- and upper-middle-income countries (Appendix Figure E.7).

IEG found lower work quality ratings—particularly screening, appraisal, and structuring—for innovation and entrepreneurship projects than for the rest of the IFC portfolio evaluated during this period. Previous IEG evaluations have found a strong association between project performance and IFC work quality ratings (IEG 2010). Such relationships may help explain the difference in project performance between innovation-related projects and the rest of the IFC portfolio. Given that IEG did not disaggregate projects according to their innovation component, the impact of a project's innovation component on overall performance cannot be assessed. The average financial rate of return and economic rate of return for innovation projects were above the IFC benchmark of 10 percent, indicating that these investments were profitable to financiers and contributed to economic growth in the countries where they were implemented. The average financial and economic rates of return for innovation-related projects were not statistically different from other projects, indicating that innovation projects performed just as well as others in their contribution to firm profitability and welfare of society (Appendix Table E.10).

In IFC's innovation and entrepreneurship projects, issues in three areas accounted for the majority of problems associated with partly unsuccessful or lower outcomes sponsors, markets, and risk (Table 4.3). The first two issues relate to IFC's front-end work, whereas supervision is mainly an implementation issue. Given the high risks associated with innovation-related projects, there is likelihood that IFC may have identified these issues but

BOX 4.4 Business Incubators in Armenia

A shortage of well-educated and qualified human capital poses an important constraint to firm-level innovation and competitiveness. The World Bank supported the creation of the Enterprise Incubator Fund (EIF) in the early 2000s in Armenia. The project had three main components: (i) managed workspace, including provision of a telecommunication infrastructure and office space to lease to interested information technology companies; (ii) a business services center providing business development services such as management and marketing skills and enabling business training and connections; and (iii) a skills development facility providing training for enterprise incubator tenants and students.

IEG's case study assessed the benefits and challenges of the business incubator from the perspective of beneficiary and non-beneficiary firms. It involved a random sample of 49 information technology firms, of which 34 identified themselves as beneficiaries and 15 as non-beneficiaries of EIF. Beneficiary and non-beneficiary companies were relatively similar in terms of employment, employment growth, and sales/profit share. Non-beneficiary firms were more export oriented and less likely to have R&D than beneficiaries.

The two most important reasons that beneficiaries sought support from EIF were to improve their business operations and to develop staff skills and capacity. Beneficiary firms were generally satisfied with the services they received from EIF; about 55 percent derived large or very large benefits from the service. They also considered EIF training sessions as useful. The Fund also helped improve the business environment: about 63 percent of the firms reported an increase in the number of employees during the past three years; about 76 percent reported annual growth in sales revenues and/or profits as well, which averaged 14.3 percent. About 63 percent of the surveyed firms successfully exported their products and services; about 57 percent reported that their businesses would be affected negatively if they stopped getting services from EIF. About 43 percent of beneficiaries plan to continue receiving services from EIF.

Respondents had varied perspectives on the role of the EIF and the development of the information technology sector. More beneficiaries (56 percent) than non-beneficiaries (40 percent) believed that EIF promoted the interests of information technology enterprises and the sector. As well, more beneficiaries (32 percent) than non-beneficiaries (13 percent) believed that EIF helped to promote innovative ideas. In general, beneficiaries were more likely to acknowledge the importance of EIF.

SOURCE: IEG field studies.

TABLE 4.3 Factors Associated with Project Performance in IFC Investment Projects

Factor	Competitiveness Not Reached at Firm and/or Sector Level		Competitiveness at Firm and/or Sector Level		Ratio
	N	%	N	%	
Inadequate sponsor assessment[a]	54	48	5	5	9:1
Inadequate market assessment[a]	68	60	8	8	7:1
Inadequate risk assessment[a]	73	65	10	10	6:1
High-risk project	56	55	36	41	1:1
High-risk sponsor[b]	52	51	27	31	2:1
High-risk market[b]	77	75	54	61	1:1
Total	113		98		

SOURCE: IEG.

NOTE: Risk assessment (that is, project risk, sponsor risk, and market risk) was available for 102 projects that did not contribute to competitiveness and 88 projects that contributed to the competitiveness at firm or sector level. Therefore, the relevant calculations are based on the risk numbers.

[a] Information is from IFC's work quality assessment.

[b] Information is from IEG's risk database.

underestimated their implications on development outcomes. Implementation setbacks were encountered in projects regardless of their development outcome ratings. On a portfolio basis, the average financial and economic rate of return on innovation-related projects performed just as well as projects without innovation components.

A multivariate analysis of the drivers of outcomes in innovation-related projects found that adequate sponsor assessment and front-end market assessment had a strong and positive impact on achieving an investment project's development outcome. The quality of supervision, IFC's role and contribution, and sponsor assessment also had a positive and significant impact on development outcomes, raising the probability of achieving it by about 30 percent. Considering interaction effects, the regression analysis showed that inadequate market assessment had a negative impact on the probability of achieving project development outcomes, when projects had good supervision. These findings suggest that sound market analysis is critical in ensuring that IFC's investment projects with innovation and entrepreneurship components are effective in achieving their development outcomes (Appendix F).

Performance of IFC's Targeted Interventions

In strengthening entrepreneurial capabilities, firm-level upgrading interventions were mostly expansion projects in sectors and activities where IFC has a track record and has accumulated extensive experience. For example, in technology-upgrading interventions, IFC has traditionally supported both firm expansion—through the purchase of new technology and equipment—and upgrades to business processes and firms introducing new products or services in markets. These interventions tend to be dominated by manufacturing and financial markets, sectors where IFC has significant expertise and experience.

Across innovation and entrepreneurship-related projects, IFC's support for investment in technology upgrading through technology transfer, diffusion, or technology acquisition had the highest proportion of projects with successful or better ratings for development outcomes, returns to IFC, and PSD. In contrast, financing schemes that supported early-stage start-ups, R&D for firm-level capacity, and establishment of financial institutions had a relatively low proportion of projects with successful or better ratings on these key performance indicators when compared to the cohort of innovation-related projects (Appendix Table E.13). IEG's analysis of performance in these activities is based on relatively small samples; therefore, these findings are not conclusive and must be interpreted with caution. They may, however, be indicative of broader performance drivers and obstacles that may warrant further investigation.

Project, sponsor, and market risk also played an important role in the performance of the IFC investment projects. Some of the innovation and entrepreneurship projects had greater exposure to project-type risks than other projects because they tend to be new projects, involve the establishment of new institutions, or have new sponsors. In other cases, such as support for R&D capacity and venture capital funds, interventions mainly involved start-ups and innovative young firms that may not have a track record in the industry (Appendix Table E.12).

Other interventions such as support for R&D product development, establishment of financial institutions, and financial support for early-stage start-ups had much riskier profiles. Market risk was high for venture capital financing for start-ups, establishment of financial institutions, and firm-level R&D (Appendix Table E.12). This reflected the high level of market uncertainty that is inherent in entrepreneurial activities that focus on development and introduction of new products and services in markets. IFC support to upgrade existing technologies and processes in firms, acquire new technology, and introduce new products has been tried and tested in other markets and entails relatively low market risks, because such projects have some proof of business model or product acceptance.

STRENGTHENING ENTREPRENEURIAL CAPABILITIES

Based on development impact, firm profitability, and PSD outcome indicators, IFC's performance was strongest in its traditional areas of intervention—supporting firm-level modernization and upgrading through support for upgrading existing technologies and processes, acquisition of new technology and know-how through technology transfer or diffusion, and introduction of new products into markets. Performance was weaker in its support for the establishment of new institutions and firm-level R&D.

One of the mechanisms that IFC uses to build entrepreneurial capabilities is advisory services. IEG validated 275 Advisory Services projects between 2008 and 2010. Of the 275 completed projects, 58 supported innovation and entrepreneurship activities. Nearly all the projects were in the Sustainable Business Advisory and Access to Finance business lines. These projects mainly provided capacity building and technical assistance to entrepreneurs, introduced quality certifications, or supported the establishment of new products or services in the market. For example, IFC supported one information technology start-up company that provided services and new products to the poor in rural India. Another Advisory Services project supported enterprise development by introducing a new poultry layer business in an area where private sector enterprises were virtually nonexistent.

IFC's development effectiveness was successful in about half of the innovation and entrepreneurship projects reviewed. This performance was slightly lower than the rest of the Advisory Services portfolio evaluated during the same period, but the difference was not statistically significant. Difference in performance of innovation and entrepreneurship projects was attributed to flaws in project design, changing market contexts, and risks inherent in innovation projects. For example, in one case, the first juice company in the world to implement a GLOBALGAP certification for raw materials did not replicate the program because of its high cost.

FINANCING SCHEMES FOR EARLY-STAGE START-UPS

Experience shows that start-up financing through venture capital funds is a high-risk, high-reward venture that requires much more than financing (Goldberg and others 2011). Outcomes are highly uncertain and successful outcomes are more likely when there are technically experienced investors available to play an active role in providing technical advice and oversight when needed.

IFC invests in companies at various stages of development, directly and through venture capital funds. IEG's assessment of the development impact of financing schemes is based on an analysis of evidence from Expanded Project Supervision Reports and Evaluative Notes for 10 venture capital funds in the IFC portfolio that focused on early-stage start-ups.[7] Development outcome was rated successful or better in only 1 of the 10 venture capital interventions. The rating for business success was successful or better in 2 of the 10 cases, and PSD in 3 of the 10 cases.[8] Such performance is not out of the ordinary for these types of interventions. In general, venture capital investments are characterized by high rates of failure (over 50 percent) and low probability of generating financial returns.[9] The risk profile of such funds is typically one in which the venture capital fund aims to earn high returns from 1 or 2 of the 10 investments made (Goldberg and others 2011). IEG's case study of an IFC-supported venture capital fund in South Africa illustrated some of the challenges—based on from the firm and stakeholder perspective—facing such investments in a developing country context (Box 4.5).

Another IEG case study of an IFC-supported venture capital fund in China illustrates how close investor engagement resulted in successful outcomes for investee companies (Box 4.6).

Recognizing the importance of early-stage financing to help entrepreneurs bring innovations to market, some countries and developing agencies are devising alternative mechanisms to address the financing challenges faced by entrepreneurs. For example, the United States has developed the Small Business Innovation Research Program, a public-private partnership, to encourage small business develop new processes and products to market (Wessner 2008). The best practices from this program are being replicated by other countries seeking alternative ways to help entrepreneurs bring innovations to market.[10]

South Africa's biosciences market is far from mature. In 2001, IFC invested in the country's first biotechnology venture capital fund. Four organizations invested in the fund, committing a total of approximately $12 million. IFC's investment was one-quarter of that. The fund had a seven-year term and was eligible for two one-year extensions.

Eight proposals were chosen from more than 300. The primary return strategy for the investors was via "exits," where investments are sold off, hopefully at significant profit. The key exit strategy was to develop intellectual property of significant global value that would attract major companies from the developed world as buyers. The fund's return on investment has a profile typical of venture capital funds: three companies had zero returns, two had returns between one and two-and-a-half times, and two exits were still being negotiated in 2012. The eighth company realized a high return but was sold before the investment term was over.

By 2005 numerous changes in the environment had occurred that posed challenges, including a dramatic change in the investment environment, investing company buyouts, and lack of government support. Still, in gauging the impact of the project, there are three major findings.

First, the fund had a strong impact in terms of assisting with corporate governance, where most companies had little experience. Connections to other service providers were also enabled, and the beneficiaries stated that the fund was particularly helpful regarding funding and intellectual property issues. Second, four of the eight companies would not have been created without the support of the fund; one of these companies actually had the highest return of all the beneficiaries.

Third, despite an excellent start, the fund was too small to support the level of growth most of the beneficiary companies were capable of attaining. It lacked scale on several fronts: funding, infrastructure—particularly staffing—and investment. Because of these constraints, fund staff believed that the fund could not support the most promising beneficiaries to the next level, where they might have attained significant returns. One factor in this challenge could be the general lack of understanding of venture capital in South Africa, particularly for biosciences ventures.

SOURCE: IEG field studies.

BOX 4.6 IFC Support for Innovation through Venture Capital in China

In the mid-2000s, China saw marked improvement in its economic environment for innovation and entrepreneurial development. As a result, there was a surge in the amount of venture capital raised and the number of fund managers active in the market. However, it was challenging for small, early-stage companies to get much-needed risk capital financing.

To help address this constraint, in 2006 IFC made an equity investment of $20 million in a 10-year closed-end venture capital fund. The fund was sponsored by a China-focused private equity fund management group.

The fund had total committed capital of $210 million. It invested in 20 venture growth companies at a total cost of $163 million. About 40 percent of the investment went to the technology, media, and telecommunications sector; 30 percent to consumer goods and services; and 15 percent to both clean technology and health care. It provided long-term support to investee companies and follow-on investments to investee companies; 11 of 16 investee companies have received such investments. The fund also played a hands-on role in providing management assistance, such as strategy and operational guidance, to the entrepreneurs and in helping them manage their growth into profitable and sustainable enterprises.

Most of the fund's investee companies were innovative early-stage SMEs. For example, one company was one of the first business processes outsourcing providers focused on the banking, financial services, and insurance industries in China. The fund helped address some of the biggest challenges the company faced, such as lack of funding and fear of losing control as the company grew. The company stated that the fund's reputation and endorsement helped it gain acceptance by its clients. The fund actively assisted the company in merger and acquisitions and equity financing events and provided strategic advice on managing growth and market positioning. The company continues to hold a leadership position in the market today, enjoying a dominant market share. Although it is still in a net loss, its staff grew from 350 in 2006 to 2,737 in 2010. Company revenue grew from $927,000 to about $15 million. To explore new growth potential, the company has been active in investing further in new business and new technology.

Another good example is a firm founded in Beijing in 2005 that focused on high-quality video on demand and movie services via the Internet. The fund invested about $4 million in late 2007 and again in 2011. The company reported that it valued the fund's professionalism, commitment, and track record in successfully providing managerial guidance. The fund helped the company in its follow-on fundraising efforts

continued on page 92

and worked with management to explore business models and define strategies. Since the initial investment, the company has achieved remarkable growth. Its revenue—less than $130 thousand in 2007—reached about $16 million in the first half of 2012 and was expected be around $48–$56 million in 2012. The number of active daily users grew from 0.8 to 1 million in 2007 and to 15.8 million by mid-2012. The number of employees increased from 35 in 2007 to more than 600 in 2012. The company has gained high market acceptance and demonstrated strong performance.

From the portfolio perspective, the fund's performance has also been satisfactory. The gross internal rate of return for the fund's portfolio stands at 16 percent, with net internal rate of return at 26 percent. Portfolio companies are performing on track, with more than half experiencing positive growth in revenue and productivity. The fund manager has raised a much bigger follow-on fund of $500 million, with higher participation from commercial investors, in the absence of IFC support.

Through the investment in this fund, IFC helped an indigenous fund manager raise its first venture capital fund. This helped catalyze the development of China's venture capital industry by demonstrating that a venture capital fund can be successfully run by a local manager. Since the launch of the fund, the venture capital industry in China has seen enormous growth, driven mainly by the growth of domestic venture capital funds.

SOURCES: Company interviews and IFC project documents.

Performance of MIGA Projects

Of the 48 MIGA projects that IEG evaluated between FY04 and FY12, 18 supported innovation and entrepreneurship.[11] Even though these projects do not provide a robust sample for statistical analysis, important insights can be gleaned from their performance. Project performance is assessed on the basis of four outcome indicators: development outcome, business performance, economic sustainability, and PSD.

Nine of the 18 evaluated projects had development outcome ratings that were satisfactory or better. A similar number of projects had business performance ratings that were satisfactory or better but a higher proportion—14 of 18 projects—had satisfactory or better PSD effects. This difference between business performance and PSD effects suggests that some innovation and entrepreneurship projects may provide low returns to their financiers but may have strong positive effects on the growth of the private sector, mainly through linkages, demonstration, and technology and knowledge spillover. Twelve of the 18 evaluated projects had satisfactory or better economic sustainability ratings, indicating positive welfare effects of these projects on society and stakeholders, such as customers, suppliers, and workers. The overall development outcome and its component ratings were not statistically different from those for the 48 evaluated projects, suggesting that the performance of innovation and entrepreneurship projects is no different than that of other MIGA projects.

A project's financial success, as reflected in business performance, and its effects on the private sector, as reflected in PSD outcome ratings, are used to explore the firm-level effects of MIGA's interventions that support innovation and entrepreneurship. PSD ratings provide insights into firm-level innovative activities, such as transfer of technology and skills leading to new or improved products and processes, first-of-their-kind products and processes in the markets, and demonstration effects that are copied by other firms. All the projects that introduced new products, processes, or services into markets had successful PSD effects. This suggests that MIGA's support for introducing innovations into markets has much broader effects on the growth of the private sector and development impact through forward and backward linkages, technology and knowledge spillovers to other firms and sectors, and demonstration effects.

Even though the sample of 18 projects is too small to draw stronger statistical inferences, an analysis of development outcome indicators by types of interventions provides some interesting insights. For example, four of the five projects that supported firm-level upgrading through the introduction of new products and processes into markets had development outcome ratings that were satisfactory or better. In contrast, four of nine technology transfer projects and one of four projects supporting acquisition of new technology had development outcome ratings that were satisfactory or better.

IEG found many cases where MIGA's support for firm-level technology upgrading through technology transfer, technology diffusion, and acquisition of new technology helped promote innovation, skill development, and growth of the private sector. For example, a MIGA-supported power project in Sri Lanka provided guarantees for the country's first independent power producer. This project helped increase the supply of electricity, promoted skills upgrading through an apprenticeship program, and had spillover benefits through linkages to suppliers and consumers who benefited from more reliable electricity suppliers. Demonstration effects were also created through the signal sent by the government's support for private sector participation in the project, leading to further privatization of the power sector.

In other cases, MIGA support for technology upgrading supported South-South technology transfer and knowledge flows with important development outcomes. In Nigeria, for example, MIGA's insurance coverage for two companies introduced new technology that resulted in high-quality roofing products for consumers at prices that in turn helped increase the competitiveness in the market for roofing materials. By demonstrating the importance of quality for consumers, this intervention motivated the government to draft and implement new quality standards for roofing products in the country. Some other technology upgrading projects supported by MIGA guarantees yielded significant social benefits. For example, a project in Brazil sought to upgrade a transmission and distribution system for electricity services. Among other effects, this project was instrumental in providing access to electricity to most customer groups, including low-income households.

MIGA's effectiveness is driven by its strategic relevance, role and contribution, front-end work in assessment and underwriting, and monitoring during project implementation. The performance ratings for MIGA's overall effectiveness by different types of intervention closely mirrored development outcome ratings. By intervention types, four of the five supporting introduction of innovation into markets, seven of nine technology transfer, and one of four acquisition of new technology had satisfactory or better outcome ratings for MIGA's overall effectiveness. The quality of MIGA's assessment, underwriting, and monitoring had the lowest successful ratings, with only 7 of 18 projects achieving satisfactory or better outcomes. Assessment and underwriting occur at the front-end stages of a project. MIGA monitors the environmental and social aspects of its projects, but it does not typically monitor other aspects of project performance that relate to business performance, growth of the private sector, or broader welfare impacts on society. The quality of MIGA front-end assessment and underwriting work becomes vital, because it is through it that MIGA can have its greatest influence on a project's success. Thus, the effectiveness of MIGA's interventions to support innovation and entrepreneurship will be enhanced with improvements in the quality of its front-end work in assessment, underwriting, and monitoring.

Country Perspectives: Performance of Innovation and Entrepreneurship Interventions

The analysis of achievements in CASs provided here is related mainly to Bank Group support for R&D, strengthening entrepreneurial capabilities and linkages. The CASs did not provide information on achievements relating to financing schemes for entrepreneurs.

In Chile, the Millennium Science Initiative Project helped improve the quality of scientific research and advanced training. The successful project outcomes led to follow-on activities in the Science for the Knowledge Economy Project that established a strong and coherent policy framework, promoted high-quality S&T, and supported interaction between the public and private sectors. The strong public-private research linkages that resulted from these investments have improved the Chilean S&T system and stimulated cross-sector cooperation between research and industry. The cumulative effects of these interventions made important contributions to enhancing the effectiveness of the innovation system in Chile. Project outputs helped increase the stock of human capital in the S&T sector, raise awareness of innovation, and inform the design of an innovation strategy and policy. However, policies were not enacted that could improve the innovation systems, such as increasing S&T expenditures to desirable levels that could foster innovation.

In China, World Bank support to help promote the country's knowledge economy by improving R&D and information flows and bridging the digital divide provided important inputs that informed high-level policy dialogue on digital divide issues and the national ICT strategy. In India, agricultural and water technologies were successfully transferred to farmers through extension services. Some projects demonstrated the use of innovations such as the use of GPS technology to target production impacts and improve equity.

In Brazil, the Bank supported approval of the country's innovation law and the development of a legal framework to directly subsidize private sector R&D. Although there is little information on university-industry linkages, patenting activity increased substantially after approval of the law. Experience from Brazil also indicates that innovative and successful practices relating to the modernization of S&T policies nurtured at the state level can spill across states to generate a virtuous circle of change.

STRENGTHENING ENTREPRENEURIAL CAPABILITIES AND LINKAGES

In China, the World Bank supported firms in developing and promoting greater use of renewable energy and cleaner fuels, energy efficiency, and expanded air pollution control technologies. Its support for technology transfer promoted greater use of renewable energy and cleaner fuels, energy efficiency, and expanded air pollution control technologies. IFC investments in high-tech start-up companies and entrepreneurial SMEs also contributed to strengthening firms' capabilities through technology transfer and introduction of new products, processes, and services. In Gujarat, India, an IFC project helped introduce resource-efficient technologies, such as roof-top solar projects that have been replicated in other cities. Investment in solar manufacturing has also enabled the development of renewable energy practices by demonstrating viable, innovative business models.

Summary

In general, World Bank Group projects supporting innovation and entrepreneurship have performed just as well as other Bank Group projects. Differences in country contexts affect the performance of Bank projects but not IFC projects. A variety of mechanisms have been used to implement innovation and entrepreneurship interventions, some more effective than others. Across the Bank Group, interventions are more likely to have better performance in areas where each institution has focused its activities—the Bank in support for R&D, and IFC and MIGA in technology upgrading through transfer, diffusion, and acquisition of new technologies and introduction of proven products into markets. Interventions supported by the World Bank and IFC have had mixed performance in high-risk, high-reward areas such as venture capital funds. These tend to be areas where start-ups and high-growth SMEs dominate. At the country level, the Bank Group has impressive achievements supporting R&D and strengthening entrepreneurial capabilities, particularly where these interventions are critical in the delivery of country strategic priorities.

Endnotes

[1] Project outcome ratings comprise relevance of objectives, relevance of objectives and design, efficacy (whether project objectives were achieved), and efficiency (whether costs involved in achieving project objectives were reasonable in comparison with both benefits and recognized norms).

[2] The difference in project performance between innovation and entrepreneurship projects and other World Bank projects does not account for the fact that some innovation projects, especially those involving R&D-driven innovation, take a long time to yield results. Therefore, the effects on project outcomes are mainly felt in the medium and long term as firms build their innovation capacity. Timing is therefore important in considering the outcome from innovation-related interventions, with high risks of underestimating actual project outcomes in a short-term frame. A robust M&E system must consider this temporal dimension in defining outcomes and related indicators to measure and track performance.

[3] Major and minor projects depended on the proportion of Bank cost allocated to supporting different activities, with major projects allocating 50 percent or more of Bank cost directly to innovation and entrepreneurship activities.

[4] The component ratings considered here were assigned by the Implementation Completion and Results Report teams and have not been validated by IEG.

[5] This study, even though it was referred to as an impact assessment, cannot be characterized as such because it did not rigorously identify a counterfactual situation or attribute outcomes to the incubation interventions.

[6] Development outcome is a synthesis of project performance in four dimensions—project business success, economic sustainability, environmental and social sustainability, and PSD impact. Investment outcome is an aggregate measure that assesses whether project returns were commensurate with the cost of the loans or equity investments.

[7] IEG identified 12 venture capital funds in the portfolio but only 10 of these had evaluative evidence from XPSRs and Evaluative Notes.

[8] This assessment of IFC's support for early-stage start-ups should be interpreted with caution because it is based on limited observations and available evidence contained in XPSRs. A broader assessment of the financial and development impact of funds supporting innovation in early-stage firms would look at the characteristics of different funds in terms of job creation and preconditions to function sustainably. Such an assessment is, however, outside the scope of this evaluation.

[9] See Sahlman (2010) for a discussion of these issues.

[10] See Wessner (2008) for a discussion of these issues

[11] IEG started evaluating MIGA projects in FY04.

References

Ai, Chunrong, and Edward C. Norton. 2003. "Interaction Terms in Logit and Probit Models." *Economic Letters,* 80:123–29.

Goldberg, Itzhak, John Gabriel Goddard, Smita Kuriakose, and Jean-Louis Racine. 2011. *Igniting Innovation: Rethinking the Role of Government in Emerging Europe and Central Asia.* Washington DC: World Bank.

IEG (Independent Evaluation Group). 2010. *Climate Change Phase II: The Challenge of Low-Carbon Development.* Washington, DC: World Bank.

infoDev. 2006. "Impact Assessment and Lessons Learned from infoDev's Global Network of Business Incubators." infoDev. http://infodev.caudillweb.com/en/Project.77.html.

Sahlman, William A. 2010. "Risk and Reward in Venture Capital." Industry and Background Note, *Harvard Business Review.* http://hbr.org/product/risk-and-reward-in-vernture-capital/an/811036-PDF-ENG.

Wessner, C. W. 2008. *An Assessment of the SBIR Program.* Washington, DC: National Academies Press.

5
Learning from World Bank Group Interventions

CHAPTER HIGHLIGHTS

- Project evaluations provide mixed evidence on the effectiveness of the main mechanisms that have been used to implement interventions supporting innovation and entrepreneurship.

- Bank Group staff develop and use tacit knowledge on innovation and entrepreneurship in the course of their work, but this knowledge does not adequately flow within and across the Bank Group, resulting in organizational inefficiencies and limiting the effectiveness of Bank Group support to clients.

This chapter goes beyond project performance ratings and draws lessons mainly from IEG evaluated projects, focusing on mechanisms that have been used to implement innovation and entrepreneurship interventions across the Bank Group. Lessons are organized around the main four targeted interventions that the Bank Group as a whole has used to support innovation and entrepreneurship: (i) support to R&D infrastructure; (ii) strengthening entrepreneurial capabilities; (iii) financing early-stage start-ups; and (iv) fostering linkages between innovation actors. In addition, the chapter examines the experiences from the five countries analyzed in previous chapters to illustrate the factors behind successful initiatives in public R&D systems, human capital, and research capacity that helped countries adopt, adapt, and utilize innovations. The chapter also examines the extent to which staff working on innovation and entrepreneurship share knowledge, developed in the course of their work, within and outside the Bank Group.

Support to R&D

COMPETITIVE FUNDING MECHANISMS

The World Bank has extensive experience using CRGs to support R&D interventions in agriculture. For example, CRGs have been used to support agricultural research projects in Latin America and Africa as well as agricultural sector reform in the Europe and Central Asia Region.

When CRGs have been well designed, they have brought greater accountability and transparency in the innovation process. In the projects in Latin America, the rigor and transparency of choosing proposals was given the highest rating by respondents in an IEG survey on perceptions of the competitive fund model. In the Ethiopia Agricultural Research and Training Project, the perception that the competitive grant mechanism was not transparent undermined the legitimacy and credibility of the review process. The resulting lack of confidence in the review process was partly responsible for delays in project implementation.

CRGs have been most effective when they have been linked to institutional strategy. In Armenia, the CRG was credited with introducing competition and accountability, but the scheme was not linked to institutional strategic priorities. Consequently, it was not used effectively to provide the strategic focus that was necessary to comprehensively strengthen the agricultural research system.

A strong institutional capacity to implement and manage a competitive grant program is an essential for success. CRGs made substantial contributions to agricultural innovation in Brazil and Colombia because they complemented a relatively strong public sector framework for research. However, use of the mechanism in Nicaragua and Peru faced serious implementation problems because of weak institutional capacity. Similar weaknesses were also associated with planning and organizational problems that resulted in unsatisfactory project implementation in the Small-Scale Commercial Agriculture Development Project in Bosnia and Herzegovina, the Agricultural Research and Training Project in Ethiopia, and the Horticultural Exports Promotion and Technology Transfer Project in Jordan.

CRGs have helped develop multi-stakeholder collaboration and demand-driven research systems involving end users. Yet outreach to poor people and less developed regions has been problematic because competitive projects of this sort are not effective at targeting poor people without land or other productive assets.

In Romania, where CRGs worked well, medium-scale progressive farmers with assets and information to adapt to commercial agriculture were the main beneficiaries. In Brazil, special initiatives that targeted poorer farmers and less developed regions were critical in extending benefits from the CRG scheme to the poor. In Latin America and Africa, some projects have included the private sector as partners in the implementation of CRGs, with a view to enhance their contribution to the achievement of national policy or institutional goals. However, these interventions have focused on broadening the range of service providers rather than emphasizing a dominant role for commercial firms. For example, in Latin America, the use of CRGs did not increase the role of the private sector in provision of agricultural research.

Investments in R&D produce new knowledge and technologies that enable innovative activities by farmers and other entrepreneurs. However, effective R&D that fosters innovation and entrepreneurship must be designed and implemented in ways that incorporate client demands and facilitate institutional partnerships. World Bank experience suggests that competitive research grants can be an effective mechanism for creation, application, and diffusion of knowledge and technology that enables innovation.

In sum, key lessons to enhance the effectiveness of competitive research grants include the following:

- Clear and thorough review processes are key to ensuring legitimacy and credibility of CRGs and are critical factors for the transparent transfer and utilization of scientific results to develop new products and processes that address societal needs.

- The effectiveness of CRGs is enhanced when clear links are established between the use of competitive grants and national policy goals or institutional strategic priorities to help build innovation capacities to address economic and social challenges.

- The mechanism works best where there is strong institutional capacity to acquire, adapt, and utilize innovation developed elsewhere.

A wide range of mechanisms has been effective in building S&T capacity in several innovation projects, for example, in Chile, Colombia, Ethiopia, and Ghana. The focus has, however, been on outputs—number of scientists trained, scientific papers published, and so on—and less on outcomes or positive behavioral changes that foster innovation and entrepreneurship in public research institutions and universities.

Some projects have used technical assistance and grants to strengthen linkages between the scientific community and users of research, to build private sector research capacity, and to establish outreach centers to promote interactions among R&D and private enterprises, industry, and public sector users of research findings. For example, the Chile Science for the Knowledge Economy Project aimed to strengthen linkages among the Chilean scientific community, industry, and public sector users of research findings and to build private sector research capacity. Two main factors contributed to success in the project. First was early input by the Bank, providing substantial conceptual and technical assistance in designing the project and ensuring its readiness for implementation. Second was the participation of innovation experts, with experience from similar initiatives in other parts of the world, in the preparation of the project.

In sum, key lessons from these interventions include the following:

• Capacity-building programs have been effective in training researchers and scientists, particularly when training is linked to outcomes and positive behavioral changes that foster innovation and entrepreneurship.

• Although scientists and engineers may be critical for advanced scientific research, development, and engineering, they are not the only key element of the human capital component of the innovation infrastructure. Many innovators are businessmen and entrepreneurs with little formal education, suggesting the importance of an open enabling environment with minimal or no barriers to entry.

Strengthening Entrepreneurial Capabilities

In many countries, technologies that foster innovation are developed in public research institutes and universities. Entrepreneurs, however, need incentives, institutional support, and financing to transform technologies into products and processes that improve productivity and strengthen competitiveness. Matching grants and incubators are two mechanisms that the Bank Group has used to help bring innovations to market.

MATCHING GRANTS

Matching grants, in which companies are required to match the investment, are a popular risk-sharing mechanism that the Bank and other donors have used. In the Renewable Resources Development Project in India, a matching grant successfully supported the sale and delivery of renewable energy technologies, thereby promoting the commercialization of a new technology by private developers and entrepreneurs. In another project, the Renewable Energy Development Project in China, technology improvement companies used matching grants to improve quality of and reduce costs for wind and solar technology products.

The use of matching grants has not been effective in several projects, warranting caution. In the Industrial Technology Development Project in Indonesia, the mechanism failed to improve performance of SMEs. It was also not effective in several projects in Africa (the Ghana Private Sector Development Project, the South Africa Industrial Competitiveness and Job Creation Project, the Zimbabwe Enterprise Development Project, and the Zambia Enterprise Development Project).

Experiences from these interventions suggest some requirements for improving the effectiveness of matching grants. Sound diagnostic work is critical for successful performance. In the renewable energy project in China, successful project performance was attributed to good diagnostic work that informed the design of market-driven approaches for renewable energy technologies. Matching grants worked well when they were integrated within a broader sector or national strategy. The absence of a consistent national policy to support the development of renewable energy in the project in China was related to underestimation of institutional and policy barriers, overambitious targets, and a much smaller contribution to the development of renewable energy. This shortcoming was recognized, and a policy-oriented approach was built into the design of a follow-up project on renewable energy.

Matching grants also work well when their design includes clear eligibility criteria that ensure that the right projects and right firms are selected. In the Industrial Technology Development Project in Indonesia, where the mechanism failed to improve performance of SMEs, the lack of clear eligibility criteria led to selection of beneficiaries whose needs were not consistent with the objectives of the matching grants scheme. The lack of clarity on eligibility criteria was also associated with failed matching grant schemes in Africa.

Besides design issues, the effectiveness of matching grants has been limited by implementation problems such as complex procurement and reimbursement procedures, political capture, and bureaucratic problems. Successful matching grant schemes, including those from Africa, have involved private sector players who were familiar with the industry in the review and selection process.

In sum, important lessons from Bank experience are:

- Matching grants can be used effectively to encourage innovation in areas where the returns are uncertain or risks are high. They can also be useful in providing incentives to upgrade entrepreneurial capacities in firms that otherwise would not be linked to R&D or other sources of knowledge.

- Matching grants permit greater control of the activities or actors that can be supported. In theory, they address the key rationale for the intervention, for example, the underinvestment of R&D by firms in general, or desire to promote more innovation in a particular area, such as alternative energy technology.

- There is a need for more effort to improve the effectiveness of matching grants through better selection processes; a shift of focus from risk avoidance to risk sharing; eligibility criteria that target firms that need such support; and experimentation, combined with active M&E of alternative implementation arrangements.

TECHNOLOGY TRANSFER AND DIFFUSION AND UPGRADING OF EXISTING TECHNOLOGIES AND PROCESSES

Technology transfer usually involves the movement of physical equipment, business models, or processes. For example, IFC financed a leading steel manufacturer that transferred state-of-the-art milling technology to Bulgaria, accompanied by significant knowledge transfers, and upgrading of skills and know-how of local staff. The investment led to greater competition in the sector, and the firm became more competitive, increasing its export market shares after the transaction.

In IFC's support for South-South technology transfer involving a bottler and distributor of soft drinks, the investment was profitable, but the replication of the distribution model was not as successful in Asia as it was in Africa. Another example is IFC support for a joint venture between an Italian and Ukrainian sponsor to install a business process, which resulted in transfer of knowledge and skills, demonstrated good industrial and management practices, and led to the emergence of an internationally competitive producer.

MIGA's support for technology transfer, including its South-South transactions, has had distinct development effects (IEG 2012). Its issue of political risk insurance for equity and debt has helped jump-start private sector FDI in post-conflict situations in Nicaragua and Mozambique and has helped introduce new high-quality products to consumers, such as roofing materials in Nigeria. In many of these cases, MIGA's support for technology transfer has helped build firm-level capabilities to adopt and adapt new technologies and practices, increase sector competitiveness, and encourage replication of good practices by other firms and governments.

In summary, key lessons from technology upgrading are:

• Technology transfer and diffusion is an important source for firms to tap into global knowledge and technology, but not much is known about how technologies are absorbed, assimilated, and utilized to maximize learning, innovation, and spillovers into the broader economy.

• Good understanding of local contexts, including consumer preferences and political economy, is a key factor that helps explain the likelihood of introducing innovations through technology transfer.

INTRODUCTION OF NEW PRODUCTS, PROCESSES, AND INSTITUTIONS

Projects that introduce new products into markets involve high levels of uncertainty, primarily because the products or services are untested in new markets. Another source of uncertainty resulting from introducing new products into new markets arises from inadequate assessment of barriers to market entry. IFC's support for a software development company in South Africa was not successful in part because of significant entry barriers into new export markets. The poor performance of a leasing operations project in Tanzania was also partly attributed to IFC's optimistic appraisals, which underestimated the difficulty of introducing a financial product in a new market.

Successful projects show that close supervision with clients is critical when new products are introduced into markets because it helps IFC identify potential problems at an early stage and take corrective action in a timely manner. A renewable energy project in Peru, however, associated IFC's close supervision—including the careful monitoring of subprojects and client performance—with the successful introduction of new renewable energy products into markets. Strong and long-term commitment of a technical partner has also been a key success factor for effective introduction of new products into markets, particularly when the commitment is secured through equity participation. IFC's relationship with strong technical partners was very important in the successful introduction of new banking and financial products in Afghanistan, the Democratic Republic of the Congo, Georgia, and Ukraine.

There are many similarities in IFC's experiences, from projects that introduce new products into markets and those that establish new institutions, particularly financial institutions. In addition to the challenges identified here, regulatory and legal environment and political economy considerations tend to have a large impact on the effectiveness of establishing new financial institutions. Examples from IFC-supported microfinance institutions in countries such as Benin, Mozambique, and Papua New Guinea show that lack of a conducive policy and regulatory framework, inadequate local presence, and weak institutional capacity can render interventions ineffective even where there is a strong technical partner.

Key lessons on factors that enhance the effectiveness of IFC's interventions to facilitate introduction of new products in markets and establish new financial institutions are:

- Introduction of new products, processes, and institutions involves risks because of unproven products, untested markets, and high entry costs into markets. This suggests the importance of realistically assessing upside potential, adopting a portfolio approach to expected returns, and aligning staff incentives with respect to tolerance to individual project failures.

- Up-front analytical work—by IFC Advisory Services or outside consultants—is critical for identifying new business opportunities; solid market assessment and risk analysis are also essential for dealing with uncertainties related to introducing new products into unproven markets.

- Active engagement and proactive supervision are very important in the early stages of introducing new products and financial institutions—introducing new products and institutions requires some hand-holding, active monitoring, and an ability to take corrective action in a timely manner when things go wrong.

Financing Early-Stage Start-Ups

VENTURE CAPITAL FUNDS

IFC has provided the most extensive support for venture capital within the Bank Group. Venture capital funds have been used to address a critical shortage of risk capital and have provided capital, managerial talent, and marketing support to investee companies, particularly in start-ups and SMEs with high growth potential.

IEG's analysis, based on 12 venture capital funds, 10 of which had evaluative evidence from XPSRs and Evaluative Notes, provides some useful insights. Even though many funds met their objectives of investing in their targeted investee companies, they have had low financial returns. For example, IFC's investment in a high-tech fund in India did not achieve its financial objectives five years into the project's 10-year life. In another fund in Mauritius, the internal rate of return was negative, and the fund size was one-third of the expected amount. It is important to note, however, that market practice typically assesses a fund's performance on a portfolio basis against its peers of the same vintage, rather than on stand-alone financial return at a point in time.

Despite the low returns to financiers, venture capital funds have had strong PSD effects, with good development outcomes. In the IFC-supported fund in India, investee companies pioneered new products and spinoffs into new companies. Another IFC-supported venture capital fund in Brazil implemented innovative concepts, such as hospital management services and home health care. Investments in venture capital efforts have provided companies with an alternative to loan financing and development of management skills and expertise on financial markets.

In general, venture capital funds have high risks and are often located in difficult investment environments. The examples of venture capital funds supporting early-stage start-ups that were evaluated suggest that poor performance in these funds is due to inexperienced fund management and lack of technically experienced investors who can provide technical advice and oversight when it is needed; limited viable investment opportunities; poor selection of investments; lack of multiple rounds of funding; and limited options for firm exit.

The strong PSD effect from IFC investment in venture capital funds, however, suggests that this type of financing can be an important mechanism for fostering innovation, entrepreneurship, and growth of private enterprises. Lessons identified in XPSRs suggest that IFC's interventions in venture capital funds have not been very effective. Key concerns identified in these project evaluations include IFC's relatively short-term perspective for these investments, when in reality they tend to be long-term and involve early-stage start-ups that require plenty of attention and technical support during long periods of incubation. Poor performance has been related to inadequate IFC supervision and monitoring.

In other cases, such as with technology and bioscience start-ups, venture capital funds require specialized technical expertise and understanding of the sector that IFC could not adequately provide from its own expertise and resources. This situation is aggravated when the sponsor does not have the required expertise or IFC cannot attract technical partners as co-investors or long-term consultants.

Five key lessons emerge from IFC's support for venture capital efforts:

- Start-up financing is a high-risk, high-reward enterprise, so venture capital funds need a long-term perspective before returns of investment may materialize.

- Venture capital funds' investment strategies should be robust enough to account for failed investments without jeopardizing overall fund performance.

- Good management, providing sound management and commercial expertise, is a critical factor for fund performance. Investee companies may need a lot of hand-holding and support to become viable companies that can successfully part from the portfolio company.

- Investee companies are more likely to grow when they have access to high-quality technical assistance.

- Venture capital funds are more likely to succeed where there is an adequate supply of start-ups and high-growth SMEs emerging from earlier stages in the innovation process. Start-ups and innovative SMEs that have graduated from early financing sources, such as angel investors, minigrants, and donor or government financed grants, are more likely to generate sound investments and deal flows.

Fostering Linkages among Innovation Actors

BUSINESS INCUBATORS

The World Bank and IFC have supported incubators, more recently through grants to infoDev, as a mechanism to help bring innovations to market. In Armenia, the World Bank supported the creation of the Enterprise Incubator Fund to accelerate the growth of start-ups and nascent enterprises that bring innovative technologies and services to market. The IEG country case study found that the effort supported the growth of private enterprises across many sectors, and interest was growing. A key driver of the incubator's success was supportive government policies that created an environment in which such incubators could operate successfully. However, IEG raised concerns about the sustainability of the project, noting the dependence on donor funding and the risk that implies for achieving long-term financial sustainability. An inadequate M&E system also made it difficult to assess the impact of the scheme.

An IFC-supported ICT business incubator in Senegal was not very effective in supporting the growth of innovative technology start-ups, mainly because delays in opening the incubator resulted in a smaller number of participating companies than expected. According to the Project Completion Report, none of the companies had graduated from the incubator at the time of project completion; hence, there were no data on sales or revenue from graduated companies. A study in Europe and Central Asia noted that incubators have not been very effective, either in successfully promoting businesses or in cost effectiveness (Goldberg and others 2011).

Reasons for the limited effectiveness on business incubators included the necessity for specialized skills and knowledge, supply-driven initiatives relying on high levels of subsidies, self-selection processes that tend to attract weaker companies, and concerns about financial sustainability. There is limited evaluative evidence on incubators, but their mixed performance suggests a need for caution and to prioritize impact studies to generate evidence of their development impact.

Learning and Knowledge Flows on Innovation and Entrepreneurship at the Bank Group

Lessons from operations are an important part of the knowledge agenda at the World Bank Group. In innovation and entrepreneurship, as in most thematic activity, the extraction of lessons and extent of knowledge sharing across the Bank Group and externally is essential to enhance the effectiveness and value of its support. Much of the Bank Group's knowledge is embodied in tacit knowledge, the knowledge that staff develop and use from their experience in development work. Such knowledge is often transferred through staff rotation, cross-support, or direct individual interaction (IEG 2011). Given that tacit knowledge is not easily stored or codified, IEG interviewed and surveyed team leaders with recent operations experience on innovation and entrepreneurship interventions across the World Bank Group to assess the extent of learning, knowledge flows, and channels for knowledge sharing on these issues.[1]

IEG surveyed a total of 107 team leaders—62 in the Bank and 45 with IFC. Overall, about 80 percent of staff who responded shared some experience on project design and implementation with their colleagues. Sixty-one percent of Bank staff and 31 percent of IFC staff shared information on lessons from project M&E. In general, staff were less inclined to share information on best practices that worked across several projects. Team leaders indicated that it is challenging to capture best practices because there are no mechanisms or time allocated to extract and transmit lessons from operations over time.

About half of the staff—more in the Bank than in IFC—used formal channels, such as brown bag lunches, workshops, and conferences, to share lessons from project experiences. Workshops, brown bag lunches, and conferences were used in about 70 percent of cases. Informal channels, such as peer review, coffee conversations, and other unstructured interactions with individuals, were also important for sharing information, although they were used less frequently than formal interactions (Appendix G). These patterns are consistent with observations on tacit knowledge across the Bank Group (IEG 2011).

Much of the knowledge flows on innovation and entrepreneurship happen within silos. Task leaders at the Bank and IFC shared lessons formally in 30 percent of cases with colleagues working on lending or investment operations, and with colleagues working on strategy and analytical work in 25 percent of cases (Appendix G). However, team leaders reported that most knowledge generated in a network or sector was not shared with other networks or sectors within and across the Bank Group. When lessons were shared formally, it was often within a network (cluster) rather than within the region in which they worked. Staff shared lessons with another network in about 20 percent of cases and less than 15 percent with staff in another region.

The limited flow of knowledge on innovation and entrepreneurship was more striking across the Bank Group institutions. Bank team leaders reported that they shared information on innovation and entrepreneurship formally in 8 percent of cases with IFC staff and 2 percent with MIGA staff. At the same time, Bank staff interactions with external partners on these issues occurred in about 8 percent of cases.

Country Perspectives: Opportunities, Gaps, and Lessons across Countries

Bank Group interventions have successfully supported key building blocks in developing innovation capacity in client countries. Experiences from the five countries illustrated successful initiatives in public R&D systems, human capital, and research capacity that helped countries adopt, adapt, and utilize innovations. Interventions have also helped entrepreneurs improve their business skills and capabilities so they can tap into domestic and global sources of innovation and gain access to financing for their innovative activities. In other cases, domestic institutions have been supported in the creation, application, and diffusion of knowledge to solve societal challenges.

These experiences provide useful insights into opportunities, gaps, and examples that the Bank can learn from and possibly replicate and scale up to provide solutions in these countries as well as in other countries with similar economic and institutional characteristics. The Bank's experience supporting R&D, and IFC and MIGA's experience in technology upgrading, provide good examples of opportunities, and financing of entrepreneurs and fostering linkages illustrate gaps. Lessons are bulleted under each point.

SUPPORT TO R&D

Bank support to R&D in Chile demonstrated that consistent country support and appropriate sequencing of projects and interventions is critical for developing synergies that ensure success in country innovation capacity. In this case, the Millennium Science Initiative successfully tested interventions that focused on building scientific capacity and cutting-edge research. Successful components in this project led to follow-on activities in the Science for the Knowledge Economy Project, which focused on the systemic dimensions of improving the innovation system in Chile.

- Piloting plays an important role in the innovation cycle before interventions can be successfully scaled up. Activities that support pilots and scale-up of interventions can help build innovation capacity at the country level.

China's success at using technology transfer to help firms upgrade technologies and tap into global sources of knowledge and technology demonstrates the usefulness of such strategies in promoting ideas for innovation. The learning embodied in these experiences develops new capabilities that can be replicated and scaled up to enhance South-South technology transfer and knowledge flows, as well as speed up innovation processes.

- Support for technology upgrading—technology transfer, diffusion, acquisition of technology, and introduction of new products—is critical in developing entrepreneurial capabilities in firms.

FINANCING EARLY-STAGE START-UPS

A wide range of financing mechanisms—grants, loans, venture capital, and equity—have been used to support innovation and entrepreneurship. The World Bank and IFC independently use complementary financing mechanisms to support different entrepreneurs. But there appears to be institutional specialization, with the Bank predominantly using competitive and matching grants and IFC using venture capital and loans to finance entrepreneurial activities. Notwithstanding this diverse range of financing, entrepreneurs do not benefit from complementary and sustained Bank Group financing at different stages in the innovation process.

- Effective sequencing of risk financing through a broad range of instruments would lay a solid foundation for new ways of targeting support to start-ups and other entrepreneurs with innovative ideas.

FOSTERING INNOVATION LINKAGES

Bank Group institutions have had solid experience and isolated success in supporting individual components of innovation systems—support for R&D, strengthening entrepreneurial capabilities, and financing start-ups. But these efforts rarely add up to the systemic efforts that are required to build strong linkages that encourage firms to use the products from R&D and consistently develop new products, processes, and services that solve major development challenges. Mechanisms to distill and distribute practical knowledge that foster innovation and entrepreneurship in specific country contexts are not well developed.

- Cross-sectoral and integrated solutions can help support countries in creating, diffusing, and using technology and knowledge for development solutions.

Summary

Evaluated projects show what mechanisms work, what does not work, why, and in what contexts. Some mechanisms have been effective, helping provide development solutions, and others have not. There are, however, challenges, particularly in areas where the World Bank Group does not have a long history of operations or experience using specific mechanisms. Bank Group staff develop and use tacit knowledge on innovation and entrepreneurship in the course of their work. Most staff share project experience, but the flow of knowledge on innovation and entrepreneurship is limited. The lessons from using a country lens show that there is good potential for synergies and complementarities that can enhance the effectiveness of Bank Group support to innovation and entrepreneurship in developing countries. Leveraging these benefits would, however, require a more coordinated set of actions within and across sectors and institutions of the World Bank Group.

Endnote

[1] Team leaders here refer to task team leaders at the World Bank and task managers or transaction managers at IFC.

References

Goldberg, Itzhak, John Gabriel Goddard, Smita Kuriakose, and Jean-Louis Racine. 2011. *Igniting Innovation: Rethinking the Role of Government in Emerging Europe and Central Asia.* Washington DC: World Bank.

IEG (Independent Evaluation Group). 2012. *The Matrix System at Work: An Evaluation of the World Bank's Organizational Effectiveness.* Washington, DC: World Bank.

——. 2011. *Capturing Technology for Development.* Washington, DC: World Bank.

6 Conclusions and Recommendations

There is growing recognition that innovation is critical for growth and to address urgent development challenges. In many developing countries the acquisition, transfer, and adaptation of technologies and knowledge that exist in other parts of the world is an important source of innovation-driven growth. Countries seeking to pursue such growth strategies must build their innovation capacity.

This evaluation was designed to assess World Bank Group support for innovation and entrepreneurship. The overarching question addresses the extent to which targeted Bank Group interventions foster innovation and entrepreneurship that is intended to transform new ideas into greater competitiveness, economic growth, and poverty reduction.

IEG found that Bank Group interventions have helped developing countries build their innovation capacities in different areas. However, current corporate and sector strategies do not provide adequate guidance on how to develop effective innovation interventions in a holistic manner. In fact, the World Bank Group does not have a comprehensive strategy and results framework for projects supporting innovation and entrepreneurship.

This is partly because such an agenda is still evolving. Bank Group interventions in this field have tended to be articulated around other thematic areas of interventions, and not necessarily around innovation and entrepreneurship as a theme. The policy rationale for supporting innovation and entrepreneurship projects has evolved from a narrow focus on market and government failures to a much broader perspective that considers other bottlenecks impeding innovation and entrepreneurship.

Bank-supported innovation and entrepreneurship projects perform just as well as other Bank projects. On a portfolio basis, IFC's innovation-related projects performed just as well as projects without innovation components, generating a financial and economic return that were above IFC's benchmarks. The limited evaluated projects supporting innovation and entrepreneurship performed just as well as other MIGA projects. Interventions were more likely to perform well in areas where the Bank Group has operational experience.

From a country perspective, the individual efforts for different Bank Group institutions do not address the systemic nature of innovation that is required for solving development challenges at the country level. In addition, limited mechanisms and weak incentives for learning from design and implementation restrict knowledge sharing on innovation and entrepreneurship among sectors, regions, and Bank Group institutions. There is increasing client demand for work on innovation and entrepreneurship but this needs to be better reflected and integrated across Bank Group operations and analytical work. Such efforts will help in improving the effectiveness of work on the ground and articulating a consistent set of messages to clients.

IEG proposes the following recommendations to strengthen the effectiveness of Bank Group support for innovation and entrepreneurship.

There is a myriad of activities on innovation and entrepreneurship within the Bank Group but few formal efforts to coordinate, consult, or link these activities across sectors, networks, and institutions. A well-coordinated cross-sectoral set of actions needs to emerge from different Bank Group activities on innovation and entrepreneurship. Going forward, there is need for better planning, joint decision making, improved coordination, and quality control of the Bank Group's work on innovation and entrepreneurship.

RECOMMENDATION 1: The Bank Group should develop and implement a consistent and well-coordinated strategic framework that highlights the relationships between work on innovation and entrepreneurship across different sectors and institutions. This framework should be developed, considering the context of the new Bank Group strategy and providing the building blocks for developing innovation strategies, policies, and programs that will help client countries strengthen innovation-driven growth.

• The FPD Network has an explicit practice that focuses on innovation and entrepreneurship, so it is well placed to provide the multisectoral coordination that such an effort demands.

The World Bank and IFC have provided financial support for early-stage start-ups through venture capital funds as well as loans and grants to innovative and entrepreneurial companies and SMEs. World Bank financing support for start-ups has mainly focused on matching grants and a few projects have included venture capital funds. Relative to the Bank, IFC has invested more in venture capital funds and other private equity funds that focused on early-stage and innovative firms. World Bank and IFC financing for early-stage start-ups has had mixed results and there is need for a more systematic assessment of performance drivers and obstacles. There is an urgent need to understand the conditions under which venture capital funds and other types of risk financing are likely to be successful, particularly in developing countries that have limited funding opportunities for early-stage financing. In such contexts effective support for start-ups should consider issues such as investment capital (seed capital, minigrants) at early stages of enterprise formation, weak or nonexistent markets, limited deal flows, policy dialogue, and financing regulations.

RECOMMENDATION 2: The World Bank and IFC should assess, develop, test, and learn from alternative approaches to provide risk financing for early-stage start-up firms that are at different stages of commercial growth.

- This is a fruitful area for collaboration between the World Bank and IFC, building on their respective comparative advantage.

- The discussion of financing for early-stage start-ups should not be done in isolation but embedded within an overall discussion of Bank Group support for innovation systems. Risk financing for early-stage start-ups should consider systemic and long-term conditions that are required for financing entrepreneurs in different stages of maturity within innovation systems.

The Bank Group, particularly IFC and MIGA, support technological upgrading activities in firms through technology transfer, diffusion, upgrading of technologies and processes, and introduction of new products, processes, and business models. These interventions have provided important sources of innovation in firms and countries. Such efforts need to be strengthened and made more systematic to enhance learning and knowledge flows between and across firms and countries.

RECOMMENDATION 3: The World Bank, IFC, and MIGA should take proactive steps to distill, document, and facilitate knowledge sharing on approaches to facilitating innovation from technology transfer, diffusion, and upgrading of technologies.

Much of the Bank's work on innovation and entrepreneurship is concentrated in lower- and upper-middle-income countries. But innovation is important at all stages of development, and clients from low-income countries are increasingly requesting Bank support for projects that address challenges specific to developing country contexts. Countries such as China and India have become significant actors in inclusive and incremental innovation that can be scaled up to other developing countries. Thus, there are promising opportunities to foster inclusive innovation through South-South interactions. The Bank Group needs to make special effort to develop innovation and entrepreneurship projects that address various aspects of innovation that benefit poor and other underserved populations in low- and middle-income countries.

RECOMMENDATION 4: The World Bank Group should broaden its involvement in inclusive innovation projects in response to client demands. The World Bank and IFC should intensify current efforts to pilot, assess, learn, and scale up inclusive innovation projects with partners.

- The Bank should focus on building innovation capacity early in the development process to help low-income countries acquire and adapt the types of innovation that address challenges that are specific to their local contexts.

- Teams developing inclusive innovation projects should pilot, assess, and scale up different types of inclusive innovation. Such efforts must be underpinned by an effective M&E system so that the learning process can inform dissemination and use of new products, processes, and services in other development contexts.

The World Bank Group's support to innovation and entrepreneurship has not been tracked very well. It has a diversified portfolio of activities that can provide good learning opportunities to foster innovation and entrepreneurship. However, much of the Bank's learning and knowledge is embodied as tacit knowledge that is often transferred through direct individual interaction. There is limited flow of knowledge on innovation and entrepreneurship across sectors, networks, and regions, as well as across Bank Group institutions. This leads to reliance on learning by doing, which is costly and limits effective utilization of Bank Group learning to devise efficient innovation policies and programs. The joint World Bank-OECD Innovation Policy Platform provides a mechanism that can facilitate knowledge exchange, including tacit knowledge.

RECOMMENDATION 5: Consistent with ongoing World Bank Group knowledge reform, the FPD Network at the World Bank needs to develop cost-effective and easily accessible procedures for codifying and disseminating information on project design and implementation experiences from its work on innovation and entrepreneurship. Similar efforts should be developed and implemented by IFC and at MIGA.

Project performance ratings suggest that innovation and entrepreneurship projects have mostly been successful. But there is mixed evidence on the effectiveness of key interventions and mechanisms that have been used to support innovation and entrepreneurship. There is an urgent need for more systematic assessment of innovative projects across the World Bank Group using appropriate evaluation methods. A major problem in most of the M&E information reported in project documents is that the most meaningful aspects of innovation and entrepreneurship are not measured. The few indicators reported focus mainly on R&D inputs but these do not capture innovation—new products, processes, and business models that are brought to the market. M&E of innovation policies and programs is critical to identify what kinds of policies and mechanisms are effective in specific contexts as well as to improve the efficiency of resources allocated to innovation and entrepreneurship.

RECOMMENDATION 6: A top priority is to identify innovation projects involving incubators, matching grants, venture capital, and other risk financing interventions at the World Bank and IFC that can be assessed to facilitate learning and scaling up of those that are promising.

• Teams working on innovative projects at the World Bank and IFC should build robust M&E into the design and implementation of these interventions.

Appendix A
World Bank Group Response to Market and Goverment Failures

When market failures exist, markets are not likely to provide innovation and entrepreneurship at an optimal level because the social benefits are likely to exceed the private benefits. If the private market were to provide the right level of innovation, there would be little justification for public sector involvement. If the public sector were involved, then it would displace private activity, wasting scarce public funds and effort that could be deployed elsewhere.

To complete the case for public support, it must be shown that the benefits of public interventions will exceed the costs. If a public intervention is so costly or entails public sector failures such that the costs exceed the benefits, the intervention would not raise national welfare, even if the social benefit exceeded the private benefit.

Fifty-six World Bank projects—about half of all innovation and entrepreneurship projects reviewed—explicitly identified correcting some type of market or government failure as the main justification for World Bank support (Figures A.1 and A.2). Projects typically address more than one failure, with 83 distinct market and government failures identified in the 56 projects. Bank interventions addressed four main categories of market or government failures: lack of supporting public services, incentive problems, information asymmetry, and poor business enabling environment (Table A.1). Of these four, lack of supporting public services and incentive problems were the most frequently identified failures that different types of Bank interventions were designed to solve. These market and government failures varied across sectors and regions (Tables A.2–A.4).

International Finance Corporation (IFC) project justification for innovation projects that supported innovation and entrepreneurship is based on the need to address failures in the market, at the government or firm level (Appendix Table D.12). The majority of projects, 83 percent, identified a specific market failure or firm-level constraint. Six types of failures were identified in IFC projects, with credit market imperfections the most frequently cited failure. This type of market failure is caused by factors such as lack of access to long-term capital as well as underdeveloped or poorly functioning financial systems. It was also dominant across all regions and sectors.

TABLE A.1 Failures Addressed by World Bank Group Interventions

Failure	World Bank Group Response
Government Failures	
Enabling environment Restricted access to global knowledge. This includes overly restrictive trade policy, limitations on FDI, high taxes or prohibition of technology licensing agreements, foreign travel, and foreign education. A third dimension of government failure is corruption and/or government capture by the groups it is trying to support.	Support to basic education, including higher education Reform of trade policies to encourage entry of and to decrease costs of imported products or services Increased entrepreneurship and innovation to boost competition policies and regulation Overall legal and regulatory environment
S&T policies Allocating government R&D effort to the wrong areas or industries, as well as using other government innovation instruments such as subsidized loans, venture capital, procurement, and so forth or to encourage private sector R&D through grants and subsidies in the wrong areas.	Support to S&T projects
Corruption and/or government capture by the groups it is trying to support The first is a common problem in most countries. There are no clear solutions for this except for the citizens of the country to demand more accountability of their government and government officials. The second is also quite common in many countries but this is not generally the case with innovation support.	AAA

Failure	World Bank Group Response
Market Failures	
Incentive issues Innovators unable to protect their innovations from replication by others	Establishment or improvement of support of a country's intellectual property rights regime consisting of licensing agencies, patent institutes, and a general regulatory system for licensing and transferring innovation from elsewhere Subsidization of research and development activities using fiscal incentives, grants, and matching grants Helping entrepreneurs in the commercialization of their innovation
Information asymmetry Financiers unable to invest to bring innovations to the market because of lack of necessary information about potential markets Incorrect perception of risks Coordination failures when the profitability of one investment depends on an initial investment being in place	Operation or subsidization of business incubators to help start-ups Sponsorship and support of enterprises for upgrading and innovation using matching grants, competition, soft loans, skills development, product upgrading, or export promotion Financing or support of venture capital funds using loans and grants Support for enterprise upgrading and innovation by strengthening S&T information services
Lack of support for public services Inadequate public goods and services for the stimulation and absorption of innovation	Support to Public research institutions and S&T parks for basic and applied research Public research universities, particularly the science and mathematics departments and research labs Metrology, standards, and quality control infrastructure including institutions, laws, and regulation

SOURCE: IEG.
NOTE: AAA = analytic and advisory activity; FDI = foreign direct investment; R&D = research and development; S&T = science and technology.

TABLE A.2 Number of World Bank Projects Having Market Failures Identified, by Sector

Market Failure		ARD	ED	FPD
Lack of supporting public institutions	No.	14	7	8
	%	42.42	43.75	32
Incentive issues	No.	10	6	9
	%	30.3	37.5	36
Information asymmetry	No.	9	2	5
	%	27.27	12.5	20
Poor business enabling environment	No.	0	1	3
	%	0	6.25	12
Total	No.	33	16	25
	%	100	100	100

SOURCE: World Bank.

NOTE: n = 119. ARD = Agriculture Sector; ED = Education Sector; FPD = Finance and Private Sector Development Sector.

Regional Breakdown of Innovation-Related Project Rationales—Frequency and Percentage

Market Failure		AFR	EAP	ECA	LAC
Lack of supporting public institutions	No.	10	4	6	8
	%	43	36	43	31
Incentive issues	No.	5	3	4	12
	%	22	27	29	46
Information asymmetry	No.	6	4	2	4
	%	26	36	14	15
Poor business enabling environment	No.	2	0	2	2
	%	9	0	14	8
Total	No.	23	11	14	26
	%	100	100	100	100

SOURCE: World Bank.
NOTE: AFR = Africa Region; EAP = East Asia and Pacific Region; ECA = Europe and Central Asia Region; LAC = Latin America and the Caribbean Region.

FIGURE A.1 Breakdown of Innovation-Related Project Rationales

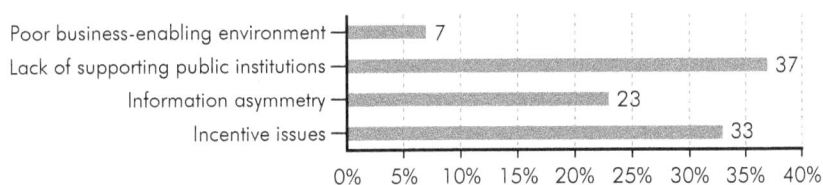

SOURCE: IEG.

FIGURE A.2 Market, Government, and Firm-Level Failures in IFC Investment Projects

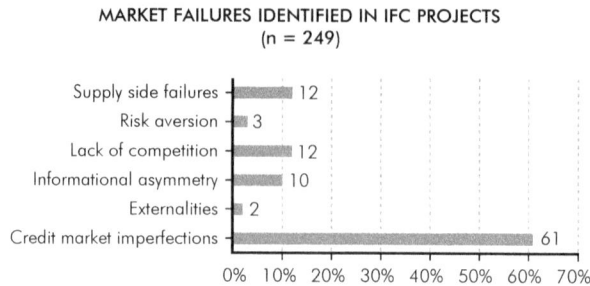

MARKET FAILURES IDENTIFIED IN IFC PROJECTS
(n = 249)

Supply side failures	12
Risk aversion	3
Lack of competition	12
Informational asymmetry	10
Externalities	2
Credit market imperfections	61

0% 10% 20% 30% 40% 50% 60% 70%

SOURCE: IEG.

The Multilateral Investment Guarantee Agency (MIGA), like IFC, aims to fix market and government failures that restrict the flow of foreign direct investment (FDI) to developing countries. FDI is viewed as an important channel for transfer of technology and as a major source of innovation in developing countries. Indeed, MIGA's Articles of Agreement stipulate that MIGA's key mandate is to support the flow of capital and technology to developing countries. By the nature of its mission, MIGA's projects aim to ameliorate market failures associated with the lack of a private source of political risk insurance. Underlying the need for political risk insurance are also government failures such as governments' inability to precommit to refrain from certain actions such as currency transfer restrictions, expropriation and breach of contract etc. that are related to political risks. Studies have found that a major concern for foreign investors is the protection of their intellectual property rights. Such concerns often result in low quality FDI or no FDI to developing countries (Smarzynska 2002). The political risk insurance industry (including MIGA) does not currently offer products that address this risk although some efforts are under way in the industry to develop it (Bullitt and Lagomarsino 2012).

TABLE A.4 Market Failures and Examples from IFC Investment Services

Market Failure	Definition	Examples
Credit market imperfection	Difficulty for firms, particularly SMEs, to raise finance for R&D, scale-up, initial commercialization, and other strategic growth issues	Lack of access to long-term capital Limited loans to SMEs Underdeveloped financial systems
Externalities	Clients and society enduring costs they did not pay for	Reduction in pollution through the development, dissemination, and/or adaptation of energy-efficiency technology
Firm capacity constraint	The firm is willing and wants to provide goods and services but does not have the capacity to do so	Lack of technology to add value to products Limited business and financial management skills
Information asymmetry	Potential financiers lack the necessary information about potential markets for a given innovation to make the necessary investment to bring it to market	Lack of early-stage funding to pursue innovations Scarcity of networks for small businesses to get access to the market
Lack of competition	The number of providers of certain goods in the market is not optimal to have competition	Monopolies Oligopolies
Risk aversion	Purchasers and professional service providers perceive high risk in dealing with SMEs	Investors refraining from investing due to the risky nature of the company Low confidence in foreign investors to provide funds due to unstable macro-economic situations
Supply-side failures	SMEs unable to respond collectively to major client requirements	Demand-supply gaps

SOURCES: IEG and IFC.
NOTE: R&D = research and development; SME = small and medium-size enterprise.

References

Bullitt, Georgia, and Laura I. Lagomarsino. 2012. "Protecting Intellectual Property Rights Abroad: New Uses for Political Risk Insurance and Standby Letters of Credit." *Berkeley Journal of International Law,* 5(2).

Smarzynska, Beata K. 2002. "Composition of Foreign Direct Investment and Protection of Intellectual Property Rights." World Bank. citeseerx.ist.psu.edu/viewdoc/download?doi=10.1.1.17.153&rep.

Appendix B
Methodology

The evaluation team identified Bank projects that supported innovation and entrepreneurship using the Independent Evaluation Group's (IEG) Implementation Completion and Results Report (ICR) database and lists of projects compiled by the Financial and Private Sector Development's (FPD) Innovation Technology and Entrepreneurship (ITE) Practice as well as an additional search of ongoing projects from the operations database. Closed projects were identified from over 2,800 ICRs completed between FY00 and FY11. IEG assembled key words that best described Bank and IFC projects and activities in innovation and entrepreneurship (see Table B.2 for a list of those words).

Next, for the World Bank, the team screened project development objectives and component information to identify projects that contained one or more of three terms: innovation, entrepreneurship, and competitiveness. Finally, the team screened these projects for words that best describe the content of Bank activities and policy instruments used to support innovation and entrepreneurship (Table B.2 lists the terms used). The team conducted a manual review of the project development objectives and component information of these projects; 64 closed projects remained in the portfolio review.[1] For active projects, IEG identified 55 relevant projects that had been identified by FPD's ITE Practice and a search of the Bank's operations database. The investment portfolio for Bank support for innovation and entrepreneurship therefore consisted of 119 projects.

For IFC investment projects, IEG followed a project selection approach consistent with guidelines of the Organisation for Economic Cooperation and Development (OECD) on collecting and interpreting innovation data for projects supporting firm-level innovation. IEG used its Expanded Project Supervision Report (XPSR) database (2000–07 calendar year approvals) and risk database (2008–10 calendar year approvals) as the main source for the portfolio review.

A three-step approach was used. In the first stage, IEG screened its database of 1,125 projects, searching project objectives, intended development impact, and IFC's role using three terms: innovation, entrepreneurship, and competition. Because not all IFC projects necessarily used these words to describe projects and activities, in the second stage, IEG included additional terms in the review of projects. These terms focused on innovation-specific instruments supporting firm-level innovation as well as innovative attributes such as new technologies, new products and services, new business models, new markets, first in country, and so forth. Finally, the team manually reviewed each project in detail, noting project objectives, intended development impact, and IFC's role to ensure that the selected projects captured attributes that describe IFC's support for innovation in its client companies. This approach resulted in a portfolio of 300 investment projects.

The project portfolio for MIGA was drawn from 371 MIGA guarantees issued between FY00 and FY12. IEG reviewed these projects in two stages. In the first stage project documents were reviewed for evidence of activities that facilitated innovation and entrepreneurship. In the second stage, a detailed desk review of each project evaluation document was used to identify projects that had innovative attributes such as new or improved ownership structures, processes, and products. IEG finally selected 108 projects that met IEG's criteria for innovation and entrepreneurship.

IEG consulted with World Bank, IFC, and MIGA staff on the methodology and criteria for selecting appropriate projects. The final list of projects selected for the study incorporates feedback and suggestions from relevant World Bank and IFC staff.

Selection of World Bank Projects

Closed Projects: To determine the list of closed Bank projects, the evaluation team utilized the ICR database. For each lending operation, Bank staff prepares an ICR shortly after the project's completion. Bank staff prepared over 2,800 ICRs between FY00 and FY11. To have a manageable group of completed projects to manually screen, the macro study team first

TABLE B.1 Attributes of Included Projects

Certification	Matching grant	Technology
Demonstration	Patent	Trade
Export	Research and development	Venture capital
Knowledge	R&D	Intellectual property
Incubator	Science	Start-up

conducted a word search in the ICR database on Project Objectives and Components. The team conducted a two-stage word screening process.

Stage 1 screened the projects for at least one of the following parts of core words to be sure to include projects with aspects in innovation, competitiveness, and entrepreneurship: "innov," "compet," and "entre."

To further narrow down the results to a manageable list, Stage 2 then screened those projects for at least one of the 17 extended attributes (Table B.1).

This two-stage process resulted in 257 *closed* Bank projects with at least one core term paired with one extended term. After manually reviewing these projects' objectives and components, the macro evaluation team selected 81 closed projects to review further for evidence of Bank support of innovation and entrepreneurship. The subsequent in-depth desk analysis resulted in selection of 64 of the 81 projects with project components promoting innovation and/or entrepreneurship.

To validate the team's screening methodology, the team applied the same two-stage word search methodology to the FPD own list of ongoing cross-sector projects under the umbrella of its recently created ITE Practice. The two-stage process successfully captured 79 percent (34/43) of the Practice's ongoing projects.

Ongoing Projects: The macro evaluation team also reviewed *ongoing* Bank projects that promote innovation and entrepreneurship. The team thus reviewed the aforementioned 55 ongoing projects in FPD's cross-sector ITE Practice. In addition, the team wanted to be certain the study captured ongoing projects supporting business incubators. Thus, an "incubator" word search in the Bank's Operations Portal revealed four more ongoing projects to include in the evaluation. IEG staff recommended an additional eight projects, some of which are included in the ITE Global Pipeline.

The final selection of World Bank support of innovation and entrepreneurship includes 119 lending projects—64 closed and 55 ongoing.

Selection of IFC Investment Projects

As IFC does not officially track innovation in its investment projects, IEG followed a selection approach consistent with the OECD's guidelines on collecting and interpreting innovation data for projects supporting innovation at the enterprise base and firm level. Applying the OECD guidelines, IEG adopted a three-step approach with a focus on terms that best capture enterprise level innovation. In the first stage, IEG screened 1,125 projects from its database (2000–11 evaluation years and 2007–10 approval years), searching project

IFC Investment Projects Extended Word Search

innovation	new delivery	science
entrepreneurship	new organization	technology
competition	new governance	knowledge transfer/know-how
new	pilot	matching grant
start-up/start-ups	new goods	venture
first	commercial	trade
research	greenfield	R&D/research and development
tech	transfer	proprietary
modernization/modern	viability	new product
improve	pioneer	new services
upgrade	latest	outsource
rehabilitation	establish/established	new business
demonstration	private equity	commercialization
patents	introduce	test
intellectual property	efficiency	productivity
certification	International Organization of Standardization (ISO)	role model
export		enterprise
new technology	replace	introduction
new business model	quality	
new marketing	seed/seed capital	

objectives, intended development impact and IFC's role for three core terms—"innovation," "entrepreneurship," and "competition." Because several IFC projects did not necessarily use these words to describe its interventions and activities, in the second stage, additional terms were included in the review of projects. These terms focused on innovation-specific instruments supporting firm-level innovation as well as innovative attributes such as new technologies, new products and services, new business models, new markets, first in country, and so forth (see Table B.2). Finally, the team decided to manually screen each project in detail reviewing the project objectives, intended development impact, and IFC's role to ensure that the right projects were capture. This approach resulted in a portfolio of 300 investment projects.

Selection of MIGA Projects

For IEG's evaluation of World Bank Group programs supporting innovation and entrepreneurship, the evaluation team focused on MIGA projects where guarantees were issued between FY00 and FY12. During this period, MIGA issued guarantees to 371 unique projects. The evaluation team manually reviewed these projects' Project Briefs for evidence of facilitating enterprise-based innovation and upgrading through (i) cross-border technology transfer; (ii) firm modernization or upgrading internally, leading to improved operations or quality of output; or (iii) the introduction of innovation into the marketplace. The team also looked for projects financing early-stage enterprises, but the review did not reveal any such projects. When the Project Briefs did not lend enough information to determine whether the project qualified for this study, the team then reviewed the relevant Board Report. After screening those 371 projects, the evaluation study team determined that 108 projects met the study's criteria for World Bank Group promotion of innovation and entrepreneurship.

Endnote

[1] This approach correctly selected 80 percent of innovation and entrepreneurship projects that were independently identified by the FPD Network.

Appendix C
Global Innovation Index Rankings for Countries in Case Study and World Bank Sector Strategies

The Global Innovation Index goes beyond traditional measures of innovation—number of research articles and the level of R&D expenditures. It measures innovation based on a combination of innovation input (that is, institutions, human capital research, infrastructure, market sophistication, and business sophistication) and innovation output (that is, knowledge and technology outputs and creative outputs) in a society.

TABLE C.1 2012 Global Innovation Index Rankings for Countries in Case Studies

Country	Income Category	Score (0–100)	Overall Rank
Singapore	High	63.5	3
Korea, Republic of	High	53.9	21
Malaysia	Upper-middle	45.9	32
China	Upper-middle	45.4	34
Chile	Upper-middle	42.7	39
Russian Federation	Upper-middle	37.9	51
Brazil	Upper-middle	36.6	58
India	Lower-middle	35.7	64
Kenya	Low	28.9	96
Rwanda	Low	27.9	102

SOURCE: Global Innovation Index, INSEAD 2012.

TABLE C.2 World Bank Group Strategy Documents Related to Innovation and Entrepreneurship

Document	Year of Strategy	Treatment of Innovation and Entrepreneurship	Degree to Which Innovation and Entrepreneurship Is Integrated into Strategy
Rural Development Strategy	2003	Among many priorities, the Bank strategy indicated it would stress "sustainable intensification through the application of science" to improve productivity. It would do this by helping developing countries access existing technologies, supporting an expansion of extension services to bring technology to farmers, and supporting commercial business development services to promote small and medium enterprises in rural areas.	Substantial
Agriculture Action Plan	2009	The first pillar of the strategy—to raise agricultural productivity—is focused on supporting adoption of technology to increase yields and improving extension services. The strategy also commits to scaling up support for a new technology generation, with more focus on regional approaches. The Bank would also support expanded rural access to financial services, which would be expected to support rural entrepreneurs.	Substantial
Education Strategy	1999	The strategy notes in passing that the context of rapid technological innovation and more exposure to global competition implies the need for a more educated workforce that can innovate continuously. There is some very modest discussion of tertiary education including with regard to science and technology, but with little emphasis or connection to innovation.	Negligible

Document	Year of Strategy	Treatment of Innovation and Entrepreneurship	Degree to Which Innovation and Entrepreneurship Is Integrated into Strategy
Education Strategy Update	2005	The update introduces a more focused concern on education-labor market linkages, advocating more systematic attention to secondary, tertiary, and science education. It acknowledges the importance of building capacity to produce and utilize knowledge. And it discusses in some detail the imperative of expanding science and technology training, and creating a network of firms, research centers, universities, and think tanks.	Substantial
Education Strategy	2011	One premise of this strategy is that learning, particularly at the secondary and tertiary level and in middle-income countries, is "critical to developing a skilled, productive, and flexible labor force and creating and applying ideas and technologies that contribute to economic growth." But there is no mention of explicit approaches to maximizing the potential to innovate through educational development. This is a presumed outcome.	Modest
Private Sector Development Strategy (including updates)	2002	The strategy considers innovation and the spread of best practice a natural outgrowth of competition, and does not offer an innovation policy beyond supporting a competitive market. IFC is said to invest in innovative projects that demonstrate the viability of types of investments and investment structures.	Modest

continued on page 136

Document	Year of Strategy	Treatment of Innovation and Entrepreneurship	Degree to Which Innovation and Entrepreneurship Is Integrated into Strategy
ICT Strategy	2012	The strategy identifies support for innovation as a strategic priority for the sector. "Innovate," one of three pillars of the 2012 strategy aims to advance ICT to improve competitiveness and accelerate innovation and target ICT skills development. The strategy articulates a vision for World Bank and IFC, working together to promote an enabling environment, support skills development and entrepreneurship, and foster innovation building on the comparative advantage of PSD, education, and ICT teams.	Substantial

SOURCE: World Bank.

NOTE: ICT = information and communications technology; IFC = International Finance Corporation; PSD = private sector development.

Reference

INSEAD. 2012. *The Global Innovation Index: Stronger Innovation Linkages for Global Growth.* Fountainbleau: INSEAD.

Appendix D
World Bank Group Project Characteristics

TABLE D.1 Project Lending of Closed and Active Projects, by Major and Minor Innovation Activities

	No. of Projects	Total Project Lending ($ millions)	Lending for Innovation Components ($ millions)
Closed			
Major	35	1,724	1,553
Minor	29	2,804	136
Closed total	64	4,528	1,806
Active			
Major	34	2,310	2,086
Minor	8	629	162
Active total	42	2,939	2,248
Total	106	7,467	4,054

SOURCE: World Bank.
NOTE: Total project lending and Innovation lending costs from closed projects is *actual* lending collected from ICRs. Total project lending and Innovation lending costs from active projects is *appraisal* lending collected from Project Appraisal Documents. Major projects are those where half or more of the project lending amount is innovation lending. Thirteen projects' lending related to innovation and entrepreneurship was not identifiable; therefore, n = 106. These were all active projects.

TABLE D.2 Lending on Innovation Component by Income Category

Income Category	Lending for Innovation Components ($ millions)	No. of Projects	Average Lending per Project ($ millions)
Lower	1,352	48	28
Lower-middle	708	36	20
Upper-middle	1,711	22	78
Total	3,771	106	36

SOURCE: World Bank.
NOTE: n = 106. Thirteen projects' lending related to innovation and entrepreneurship was not identifiable. These were all active projects.

TABLE D.3 World Bank Project Component Lending by Region

	Closed			Active		
	Lending for Innovation Components ($ millions)	No. of Projects	Average Lending per Project ($ millions)	Lending for Innovation Components ($ millions)	No. of Projects	Average Lending per Project ($ millions)
AFR	223	19	12	843	21	38
EAP	293	6	49	143	2	71
ECA	199	8	25	193	5	39
LAC	612	24	26	954	10	95
Other	196	7	28	115	3	38

SOURCE: World Bank.
NOTE: AFR = Africa Region; EAP = East Asia and Pacific Region; ECA = Europe and Central Asia Region; LAC = Latin America and the Caribbean Region. n = 106. Thirteen projects' lending related to innovation and entrepreneurship was not identifiable. These were all active projects.

TABLE D.4 World Bank Project Component Lending by Sector

	Closed			Active		
	Lending for Innovation Components ($ millions)	No. of Projects	Average Lending per Project ($ millions)	Lending for Innovation Components ($ millions)	No. of Projects	Average Lending per Project ($ millions)
ARD	520	21	25	444	11	40
ED	590	7	84	376	6	63
FPD	330	28	12	1,096	18	61
Other	83	8	10	332	7	47

SOURCE: World Bank.
NOTE: n = 106. Thirteen projects' lending related to innovation and entrepreneurship was not identifiable. These were all active projects. ARD = Agriculture Sector; ED = Education Sector; FPD = Finance and Private Sector Development Sector.

FIGURE D.1 World Bank Innovation Projects over the Years

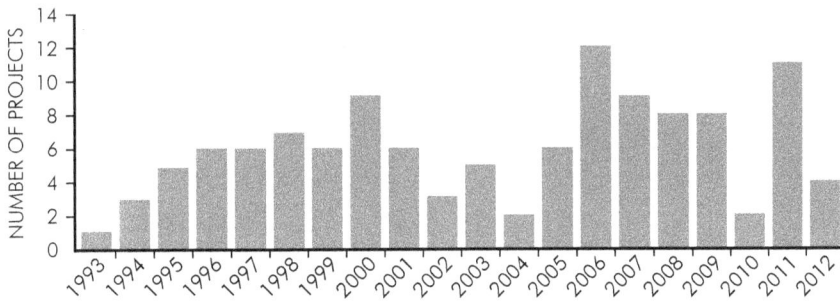

TABLE D.5 World Bank Innovation Projects Share by Sector

World Bank Sector	Share (%)	Share (volume) (%)
Agriculture and Rural Development	5	3
Education	2	3
Energy and Mining	1	1
Environment	1	0
Economic Policy	1	0
Financial Markets	0	0
Financial and Private Sector Development	11	9
Gender	0	0
Global Information/ Communications Technology	4	1
Health, Nutrition, and Population	0	0
Private Sector Development	0	0
Social Development	0	0
Social Protection	1	0
Transport	0	0
Urban Development	0	0
Water	0	0
Total	2	2

SOURCE: World Bank.

TABLE D.6 IFC Investment Portfolio Total Net Commitment in Innovation and Entrepreneurship by Project Status ($ millions)

	No. of Projects	Total Project Lending ($ millions)	Lending for Innovation Components ($ millions)
Closed	272	4,876	1,806
Open	28	833	2,248
Total	300	5,709	4,054

SOURCE: IFC.
NOTE: n = 300.

TABLE D.7 IFC Investment Portfolio Total Net Commitment on Innovation Component by Income Category

Income Category	Total Net Commitment ($ millions)	No. of Projects	Average Net Commitment per Project ($ millions)
Lower	516	69	7
Lower-middle	2,355	125	19
Upper-middle	2,356	84	28
High	102	4	25
Regional	380	18	21
Total	5,708	300	19

SOURCE: World Bank.
NOTE: n = 300.

TABLE D.8 IFC Investment Portfolio Total Net Commitment in Innovation and Entrepreneurship, by Region

Region	Closed			Open		
	Total Net Commitment ($ millions)	No. of Projects	Average Net Commitment per Project ($ millions)	Total Net Commitment ($ millions)	No. of Projects	Average Net Commitment per Project ($ millions)
SSA	240	32	8	231	3	77
LAC	1,599	63	25	155	8	19
ECA	1,364	73	19	343	12	29
MENA	222	24	9	58	2	29
EAP	928	41	23	42	2	21
SAR	515	36	14	4	1	4
World	9	3	3			
Total	4,877	272	101	833	28	179

SOURCE: IFC.
NOTE: n = 300. EAP = East Asia and Pacific; ECA = Europe and Central Asia; LAC = Latin America and the Caribbean; MENA = Middle East and North Africa; SAR = South Asia; SSA = Sub-Saharan Africa.

TABLE D.9 IFC Investment Portfolio Total Net Commitment in Innovation and Entrepreneurship, by Sector

Sector	Closed			Open		
	Total Net Commitment ($ millions)	No. of Projects	Average Net Commitment per Project ($ millions)	Total Net Commitment ($ millions)	No. of Projects	Average Net Commitment per Project ($ millions)
Agribusiness and Forestry	1,006	47	21	122	5	24
Consumer and Social Services	232	21	11	13	2	7
Financial Markets	1,106	75	15			
Funds	113	10	11			
Global Product Group	3	1	3			
Infrastructure	468	15	31	15	1	15
Manufacturing	1,397	67	21	530	16	33
Oil, Gas and Mining	58	2	29			
Telecom and Information Technology	494	34	15	153	4	38
Total	4,877	272	18	833	28	30

SOURCE: IFC.
NOTE: n = 300.

IFC Innovation Projects over the Years

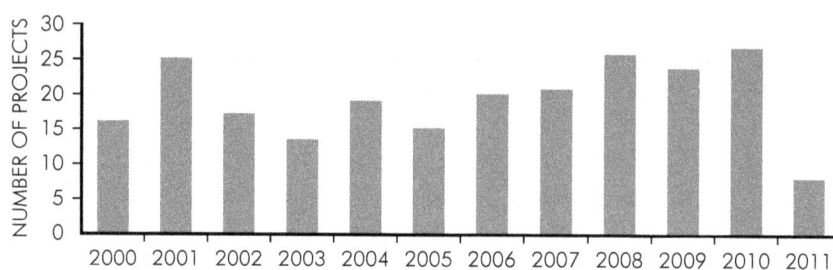

SOURCE: IEG.

TABLE D.10 IFC Innovation Projects, by Sector

IFC Sector	Share (%)	Share (volume)
Agribusiness and Forestry	34	39
Consumer and Social Services	16	12
Financial Markets	18	11
Funds	15	6
Infrastructure	11	11
Manufacturing	44	46
Oil, Gas, and Mining	3	4
Other Infrastructure Sectors	0	0
Other MAS Sectors	0	0
Telecom and Information Technology	64	56
Total	25	23

SOURCE: IEG.
NOTE: MAS = Manufacturing, agriculture, and services.

MIGA Innovation and Entrepreneurship Projects, by Region

Region	No. of Projects	MIGA Guaranteed Coverage, Cumulative ($ millions)	Average Guaranteed Coverage ($ millions)
EAP	5	159.7	31.9
ECA	38	1,357.1	35.7
LAC	17	974.3	57.3
MENA	4	89.9	22.5
SAR	8	400.7	50.1
SSA	36	1,851.4	51.4
Total	108	4,833.0	44.8

SOURCE: MIGA.
NOTE: n = 300. EAP = East Asia and Pacific; ECA = Europe and Central Asia; LAC = Latin America and the Caribbean; MENA = Middle East and North Africa; SAR = South Asia; SSA = Sub-Saharan Africa.

FIGURE D.3 AAA Networks (random sample)

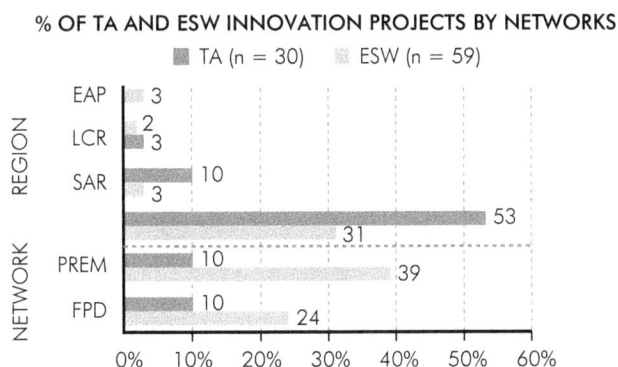

% OF TA AND ESW INNOVATION PROJECTS BY NETWORKS

TA (n = 30)   ESW (n = 59)

REGION
- EAP: 3
- LCR: TA 3, ESW 2
- SAR: TA 3, ESW 10

53
31

NETWORK
- PREM: TA 10, ESW 39
- FPD: TA 10, ESW 24

SOURCE: IEG.
NOTE: EAP = East Asia and Pacific Region; LCR = Latin America and the Caribbean Region; SAR = South Asia Region; ESW = economic and sector work; FPD = Finance and Private Sector Development Network; PREM = Poverty Reduction and Economic Management Network; TA = technical assistance.

IFC Advisory Services by Business Line

Business Line (N = 84 projects)	Number	Percent	Total Funding per Project ($)
A2F	38	45	604,672
IC	2	2	1,371,287
PPP	3	4	713,231
SBA	41	49	355,362
Total	84	100	505,115

SOURCE: IFC.
NOTE: A2F = Access to Finance; IC = Investment Climate; PPP = Public-Private Partnership; SBA = Sustainable Business Advisory.

FIGURE D.4 AAA Projects by Networks, Four Countries

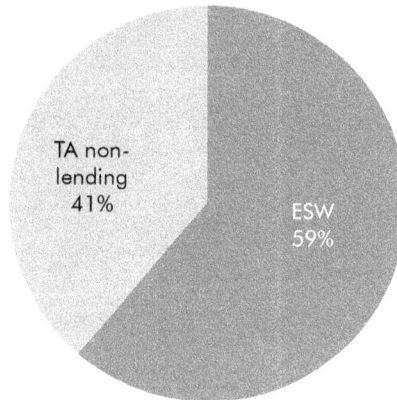

SOURCE: IEG.
NOTE: n = 71. AAA = analytic and advisory activity; ESW = economic and sector work; TA = technical assistance.

AAA Client and Strategic Tasks, Four Countries

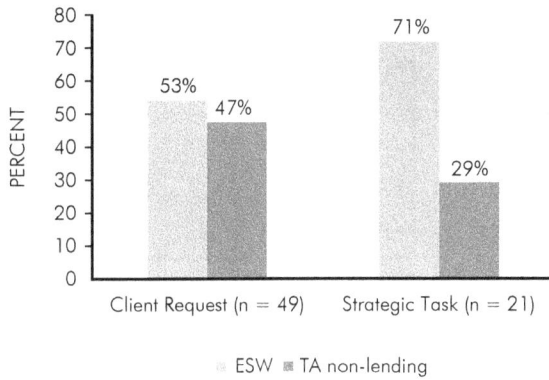

SOURCE: IEG.
NOTE: AAA = advisory and analytic activities; ESW = economic and sector work; TA = technical assistance.

AAA Sectors for Four Case Study Countries

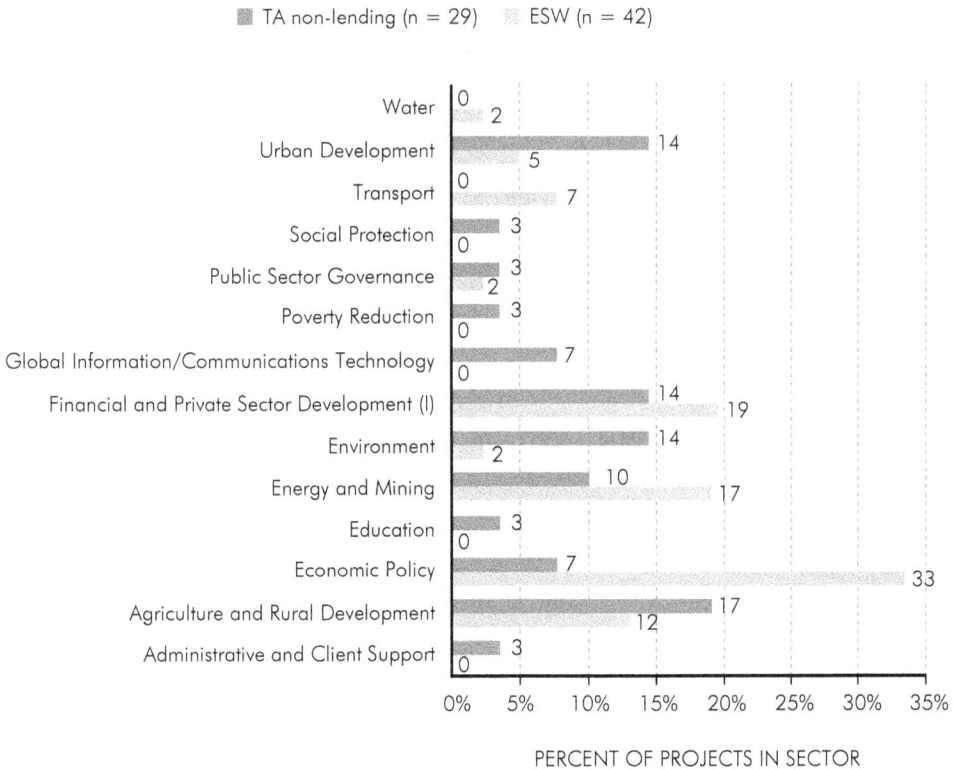

PERCENT OF PROJECTS IN SECTOR

SOURCE: IEG.
NOTE: AAA = advisory and analytic activities; ESW = economic and sector work; TA = technical assistance.

AAA Objectives (random sample)

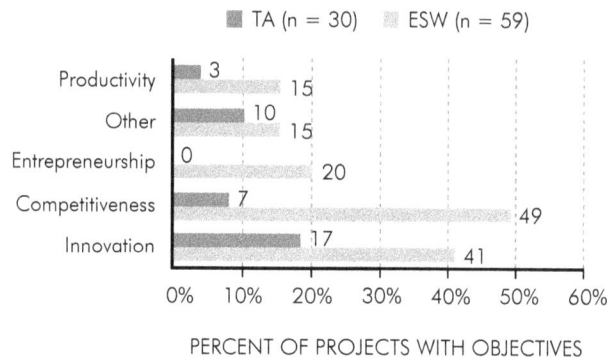

TA (n = 30)　　ESW (n = 59)

Productivity	TA: 3, ESW: 15
Other	TA: 10, ESW: 15
Entrepreneurship	TA: 0, ESW: 20
Competitiveness	TA: 7, ESW: 49
Innovation	TA: 17, ESW: 41

0%　10%　20%　30%　40%　50%　60%

PERCENT OF PROJECTS WITH OBJECTIVES

SOURCE: IEG.
NOTE: AAA = advisory and analytic activities; ESW = economic and sector work; TA = technical assistance.

BOX D.1 infoDev's Global Business Incubator Initiative

The Incubator Initiative was launched in 2002, with support from the government of Japan. Today, infoDev's global business incubator network consists of nearly 300 incubators in over 80 developing countries assisting 20,000 enterprises. infoDev supports these incubators with both financing and capacity building, and regional and global networking is facilitated through workshops and forums. In 2006, regional networks were launched with the aim of bringing together these organizations to share lessons and experiences. Through this initiative infoDev aims to—

• Provide financial and technical assistance to help existing incubators improve and scale up their operations

• Facilitate the development of new business incubators in areas that are currently not served

• Encourage innovative use of information and communications technology (ICT) based on local needs and contexts

• Enable a network of knowledge-sharing among business incubators that support ICT-enabled private sector development.

The Incubator Initiative can be viewed both as a capacity-building program and as a vehicle for research. The Incubator Initiative works towards capture and dissemination of knowledge and best practices on promoting ICT-enabled entrepreneurship. Such activities help in understanding the primary constraints to ICT-enabled innovation, new business creation and expansion across the economy. The knowledge created through the Incubator Initiative is also used to deliver customized advisory services to donors.

SOURCE: infoDev.

BOX D.2 Development Marketplace

The Bank launched the Development Marketplace in 2001. Its objective is to position social entrepreneurs as the third arm of development along with public and commercial private sectors.

The Development Marketplace targets "high social impact" entrepreneurs that need growth finance to expand, scale up, or replicate their operations in a financially sustainable manner. The Marketplace has managed a competitive grants program that helps social enterprises expand the supply of public goods and services to populations at the bottom of a developing country's income distribution. Since then, more than 300 global winners have won $200,000 each in grant funding.

In 2011, the Development Marketplace was expanded with the launch of the Development Marketplace Investment Platform to link selected entrepreneurs with capital providers. The rationale for this is that funding obtained through the Development Marketplace is essentially seed funding that mostly goes into proof-of-concept demonstrations. To leverage this seed funding, social enterprises need multiyear financing in the form of grants, loans, and equity. Investors, in contrast, are hindered by transaction costs for search and due diligence processes that tend to be high because of limitations such as readily available financial data, remoteness of the enterprise, and absence of standardized performance metrics.

The combination of the Development Marketplace and the Investment Platform addresses challenges in the social enterprise ecosystem using three stages: Development Marketplace is used to uncover innovative social enterprises; there is increased focus and commitment to the technical assistance infrastructure needed to support testing and development of these enterprises; and the Development Marketplace Investment Platform helps increase scale and impact by matching the enterprises with potential investors.

SOURCE: World Bank Institute.

Appendix E
World Bank Group Project Performance

World Bank

Innovation and Entrepreneurship versus Rest of the World Bank Portfolio Project Outcome Performance

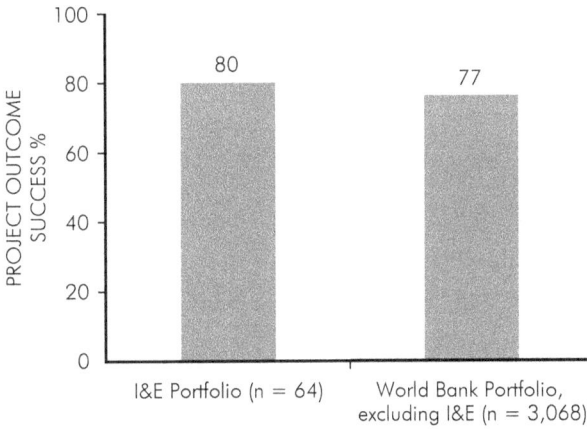

SOURCE: ICR Review database.

NOTE: The comparison is between the projects that were evaluated during the review period (2000–10). The difference between two groups is not significant.

Innovation and Entrepreneurship versus Rest of the Portfolio Project Outcome
Success, by Sector

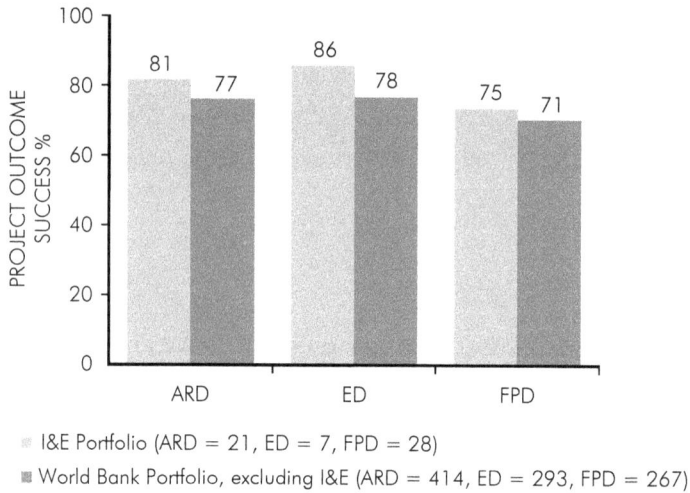

I&E Portfolio (ARD = 21, ED = 7, FPD = 28)
World Bank Portfolio, excluding I&E (ARD = 414, ED = 293, FPD = 267)

SOURCE: ICR Review database.
NOTE: ARD = agriculture and rural development; ED = education; FPD = finance and private sector development;
I&E = innovation and entrepreneurship.

Innovation and Entrepreneurship versus Rest of the Portfolio Project Outcome
Success, by Income Group

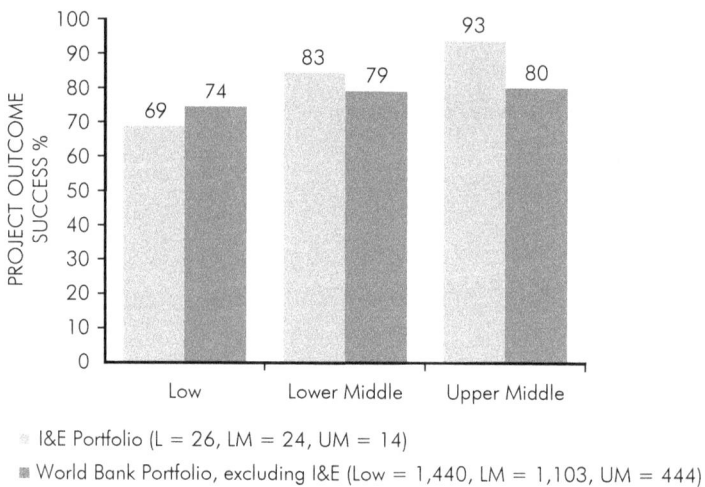

I&E Portfolio (L = 26, LM = 24, UM = 14)
World Bank Portfolio, excluding I&E (Low = 1,440, LM = 1,103, UM = 444)

SOURCE: ICR Review database.
NOTE: L = low; LM = lower middle; UM = upper middle.

Outcome Rating and Bank Performance

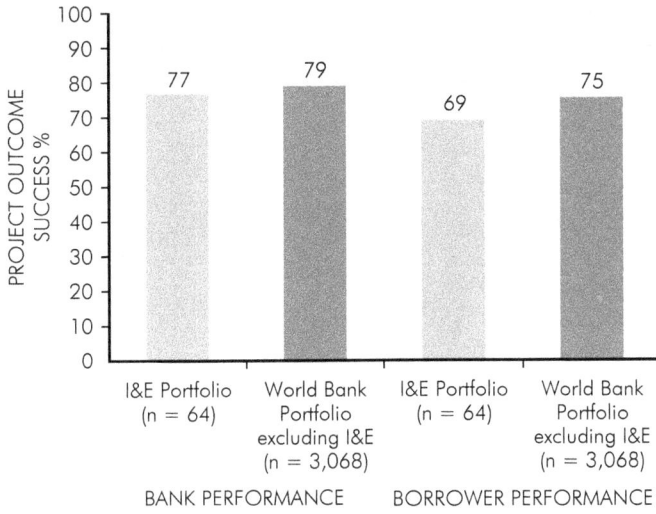

SOURCE: ICR Review database.
NOTE: I&E = innovation and entrepreneurship.

FIGURE E.5 World Bank Performance across Income Groups

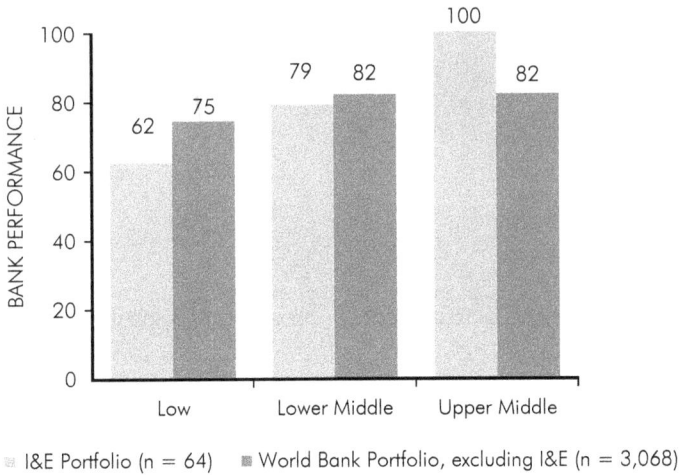

SOURCE: ICR Review database.
NOTE: I&E = innovation and entrepreneurship.

Borrower Performance across Income Groups

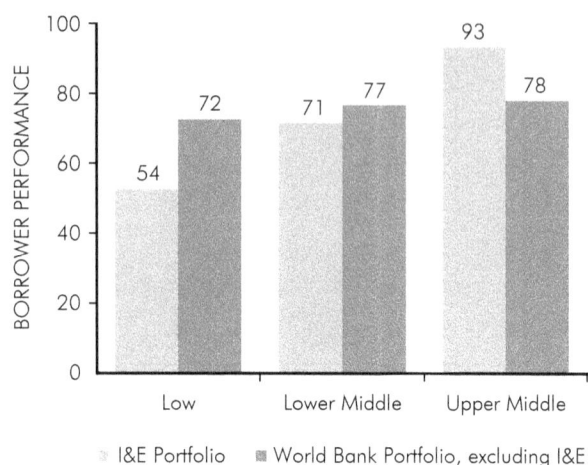

SOURCE: ICR Review database.
NOTE: I&E = innovation and entrepreneurship.

TABLE E.1 Project, Bank, and Borrower Performance by Major/Minor

	Successful Development Outcome[a] (%)	High Borrower Performance (%)	High Bank Performance (%)
Minor	69	59	72
Major	89	77	80

SOURCE: IEG.
[a] Statistically significant at p = 0.05.

TABLE E.2 Development Outcome across Income Groups and Intensity of Innovation and Entrepreneurship

High Development Outcome	Low Income (n = 13 minor, 13 major) (%)	Lower Middle (n = 11 minor, 13 major) (%)	Upper Middle (n = 5 minor, 9 major) (%)
Minor	62	73	80
Major	77	92	100

SOURCE: IEG.

TABLE E.3 Project Objective Performance by Sector

Objectives Reached	Innovation and Entrepreneurship Objectives (n = 117) (%)	Rest (n = 83) (%)
ED	62	—
ARD	63	57
FPD	53	52

SOURCE: IEG.
NOTE: n = 106. Thirteen projects' lending related to innovation and entrepreneurship was not identifiable. These were all active projects. ARD = Agriculture Sector; ED = Education Sector; FPD = Finance and Private Sector Development Sector.

TABLE E.4 Project Objectives by Sector

Objectives Reached	ARD	ED	FPD
I&E relevant (no.)	51	13	44
I&E relevant (%)	78	93	44
Other type of objectives (no.)	14	1	57
Other type of objectives (%)	22	7	56

SOURCE: IEG.
NOTE: ARD = agriculture and rural development; ED = education; FPD = finance and private sector development.

TABLE E.5 Borrower Performance across Income Groups and Intensity of Innovation and Entrepreneurship Projects

High Borrower Performance	Low Income (n = 13 minor, 13 major) (%)	Lower Middle (n = 11 minor, 13 major) (%)	Upper Middle (n =5 minor, 9 major) (%)
Minor	46	64	80
Major	62	77	100

SOURCE: IEG.

TABLE E.6 Bank Performance across Income Groups and Intensity of Innovation and Entrepreneurship Projects

High Bank Performance	Low income (n = 13 minor, 13 major) (%)	Lower middle (n = 11 minor, 13 major) (%)	Upper middle (n =5 minor, 9 major) (%)
Minor	62	73	100
Major	62	85	100

SOURCE: IEG.

TABLE E.7A I&E Projects' Bank Performance and Project Outcome Rating

Portfolio	Bank Performance	
Outcome Rating	Low	High
Low	76.92	23.08
High	9.8	90.2

SOURCE: IEG.

TABLE E.7B Rest of the Projects' Bank Performance and Project Outcome Rating

World Bank Portfolio	Bank Performance	
Outcome Rating	Low	High
Low	80.76	8.35
High	5.19	94.81

SOURCE: IEG.

TABLE E.8A I&E Project Borrower Performance and Project Outcome Rating

Portfolio	Borrower Performance	
Outcome Rating	Low	High
Low	84.62	15.38
High	17.65	82.35

SOURCE: IEG.
NOTE: I&E = innovation and entrepreneurship.

TABLE E.8B Rest of the Projects' Borrower Performance and Project Outcome Rating

World Bank Portfolio	Borrower Performance	
Outcome Rating	Low	High
Low	84.62	17.76
High	17.65	93.52

SOURCE: IEG.
NOTE: I&E = innovation and entrepreneurship.

IFC Investment

TABLE E.9 Innovation and Entrepreneurship versus Rest of the IFC Investment Portfolio Project Development Outcome Performance

	Innovation and Entrepreneurship Projects		Rest of the Portfolio	
	No. of Projects	Successful (%)	No. of Projects	Successful (%)
Development outcome	203	56	610	66
Project business success	202	47	604	55
Economic sustainability	202	57	600	70
Private sector development	202	69	604	77
Overall investment outcome	203	59	612	70

SOURCE: XPSR database.

FIGURE E.7 Project Performance by Income Group

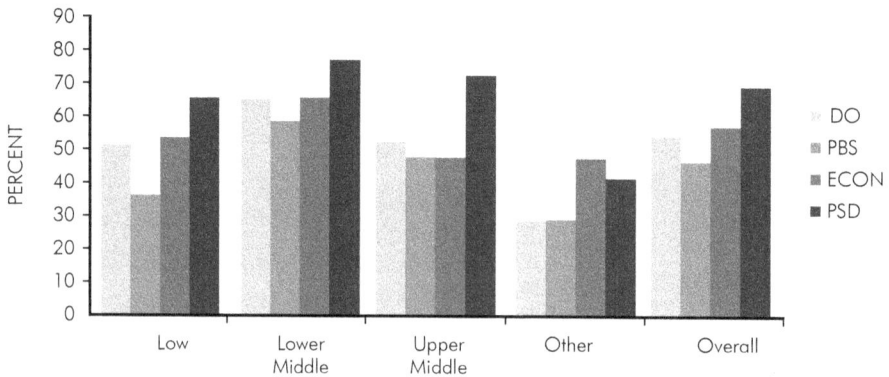

SOURCE: XPSR database.
NOTE: The figure chart shows PSD success rate minus PBS success rates. DO = development outcome; ECON = economic sustainability; PBS = project business success; PSD = private sector development.

Innovation and Entrepreneurship versus Rest of the IFC Investment Portfolio Work Quality Performance

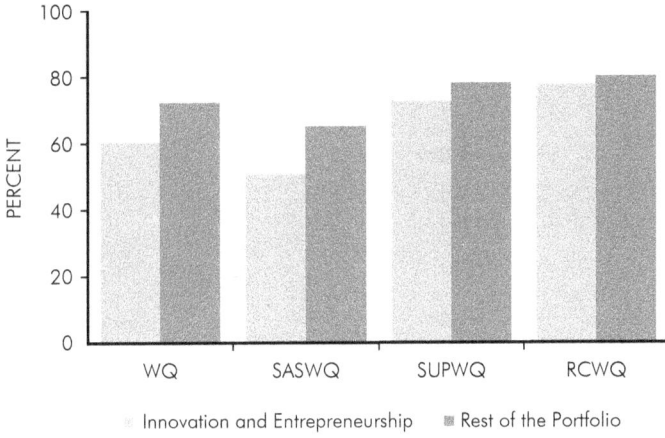

SOURCE: XPSR database.
NOTE: RCWQ = role and contribution work quality; SASWQ = screening, appraisal, and structuring work quality; SUPWQ = supervision and administration work quality.

TABLE E.10 FRR and ERR of Investment Projects

	Innovation and Entrepreneurship Projects		Rest of the Portfolio	
	No. of Projects	Average (%)	No. of Projects	Average (%)
FRR	108	12	265	14
ERR	107	17	256	18

SOURCE: XPSR Database.
NOTE: ERR = economic rate of return; FRR = financial rate of return.

IFC Investment Work Quality by Innovation and Entrepreneurship Types

	No. of Projects	SAS (%)	SUP (%)
Strengthen entrepreneurial capabilities			
Introduction of innovation into the market	41	44	63
Establishment of financial institution	28	39	61
Product or service introduction	13	54	69
Upgrading existing products and processes	70	56	79
R&D for product development	11	27	73
Technology diffusion	9	67	89
Technology transfer	55	56	58
Financing Schemes	16	25	50
Early stage financing directly to the company	6	17	67
Venture capital funding	10	30	40
Overall	202		

SOURCE: IEG.

NOTE: R&D = research and development; SAS = screening, appraisal, and structuring; SUP = supervision and administration.

TABLE E.12 Riskiness of Innovation and Entrepreneurship Investment Projects by Type of Risk

	High Project Risk (%)	High Sponsor Risk (%)	High Market Risk (%)
Strengthen entrepreneurial capabilities	47	38	67
Introduction of innovation into the market	86	47	86
Establishment of financial institution	96	54	96
Product or service introduction	60	30	60
Upgrading existing products and processes	17	35	56
R&D product development	80	70	90
Technology diffusion	50	25	63
Technology transfer	49	31	63
Financing scheme	67	80	87
Early stage financing directly to the company	60	100	100
Venture capital funding	70	70	80
Overall	49	41	69

SOURCE: IEG.
NOTE: R&D = research and development.

TABLE E.13 Project Performance of Project by Innovation and Entrepreneurship Types

	No. of Projects	Development Outcome (%)	PBS (%)	PSD (%)
Strengthen entrepreneurial capabilities	187	59	51	73
Introduction of innovation into the market	41	51	44	66
Establishment of financial institution	28	46	32	68
Product or service introduction	13	62	69	62
Upgrading existing products and processes	71	63	59	73
R&D for product development	11	27	18	45
Technology diffusion	9	56	56	78
Technology transfer	55	65	51	82
Financing schemes	16	19	6	31
Early stage financing directly to the company	6	33	17	33
Venture capital funding	10	10	0	30
Overall	203	56	47	69

SOURCE: IEG.

NOTE: PBS = project business success; PSD = private sector development.

TABLE E.14 Development Outcome and MIGA Effectiveness Rates of Success

	No. of Projects	Overall Development Outcome: Satisfactory or Better	Business Performance: Satisfactory or Better	Economic Sustainability: Satisfactory or Better	PSD Impact: Satisfactory or Better
Introduction of innovation into market	5	4	4	4	4
Technology transfer	9	4	4	7	6
Internal upgrading	4	1	1	1	3
Total	18	9	9	12	14

SOURCE: IEG.
NOTE: PSD = private sector development.

Appendix F
Factors That Play a Role in the Achievement of Project Objectives

In an attempt to better understand the factors that play a role in the achievement of project objectives in the portfolio of innovation and entrepreneurship projects, IEG identified the main factors driving success or failure from project evaluation reports.

In the World Bank portfolio of 64 closed projects for which ICRs are available, 37 projects achieved their project objectives and 27 did not. Irrespective of their success, for each of them IEG identified implementation problems, distinguishing the problems associated with the Bank role and those with the Borrower role. As shown in Figure F.1, the main setbacks within the Bank role part were mostly related to the design (complex, unrealistic, or inadequate), the monitoring and evaluation (M&E) system, and supervision. On the borrower side, the most common problems were borrower performance and implementation delays.

Two interesting features emerge from this analysis. First, both implementation problems occur both on the Bank side as well as on the borrower's side. Second, and probably more interesting, all projects are affected, one way or the other, by implementation projects.

FIGURE F.1 Distribution of Implementation Problems in Bank Group Projects on Innovation and Entrepreneurship

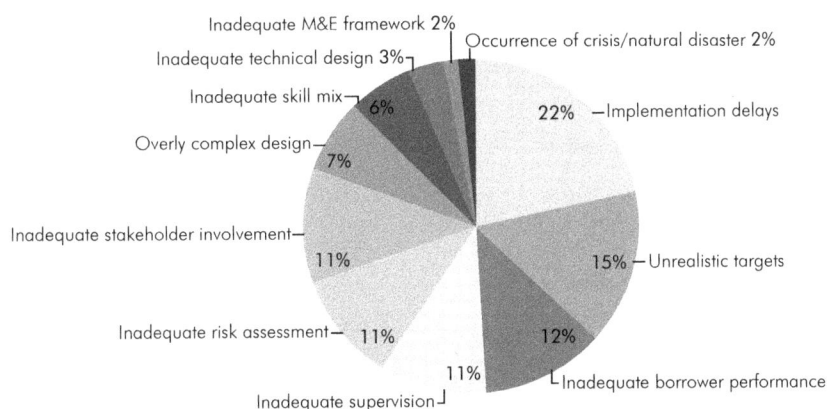

Inadequate M&E framework 2%
Occurrence of crisis/natural disaster 2%
Inadequate technical design 3%
Inadequate skill mix 6%
Overly complex design 7%
Implementation delays 22%
Inadequate stakeholder involvement
Unrealistic targets 15%
Inadequate risk assessment 11%
11%
12%
Inadequate supervision 11%
Inadequate borrower performance

SOURCE: IEG.
NOTE: M&E = monitoring and evaluation.

Setbacks occur not only in projects that do not achieve their development objectives, but also in projects that successfully achieve their goal. As shown in Table F.1, 39 percent of all problems identified by the review occur in projects that achieve their objectives

This raises a legitimate question: among all the issues identified, are some more binding than others? For instance, IEG observed that as many projects with poor M&E fail as succeeded. Similarly, inadequate technical design appears almost as often in successful projects than in unsuccessful ones. To answer this question, IEG adopts two strategies. First IEG runs a series of multivariate regressions that allow control for a number of concurrent factors, then regressions with interaction terms to establish the combined effect of problems occurring simultaneously. In all regressions, a number of fixed factors that have been shown to be

TABLE F.1 Detailed List of Problems in World Bank Projects on Innovation and Entrepreneurship

Problem	Development Objectives	
	Not Achieved	Achieved
Inadequate skill mix	3	0
Inadequate risk assessment	3	3
Inadequate technical design	19	10
Inadequate supervision	11	2
Unrealistic targets	14	9
Inadequate M&E framework	20	21
Overly complex design	14	6
Inadequate stakeholder involvement	2	1
Inadequate borrower performance	14	6
Occurrence of crisis/natural disaster	6	6
Implementation delays	10	10
Total	116	75

SOURCE: IEG.
NOTE: M&E = monitoring and evaluation.

significant in similar work were controlled for:[1] length of projects implementation (as proxy of complexity), value of project lending, sector, region, gross domestic product growth of country over the period of the project, level of economic development (lower income, low-middle, upper-middle) as proxy for institutional development, and a dummy for a project being restructured. Overall, the only variable that is consistently significant is gross domestic product growth, with has a positive effect on the probability of achieving the development objectives.

In a first set of regressions, IEG ran separately the project problems related to the Bank role and those related to the borrower behavior. These results show that after controlling for the other factors in the Bank's control, inadequate supervision reduced the probability of success by 40 percent (Table F.2, regression 1). On the borrower side, after controlling for all factors in their control, only borrower performance is significant, indicating that lack of borrower engagement reduces the probability of success by 37 percent (Table F.2, regression 2). Then a probit model is run with all projects features, both Bank and borrower. These results show that supervision loses significance. Borrower performance appears to be dominant, with the dummy for "unrealistic targeting" acquiring significance on the bank side (Table F.2, regression 3).

TABLE F.2 Probit Regression for World Bank Projects on Innovation and Entrepreneurship

Intervention	(1) Achievement of Project Objectives	(2) Achievement of Project Objectives	(3) Achievement of Project Objectives
Inadeq. risk asses. (Dummy)	0.119 (0.440)		0.103 (0.372)
Inadeq. technical Design (Dummy)	−0.245 (−1.343)		−0.267 (−1.266)
Inadeq. supervision (Dummy)	−0.411* (−1.914)		−0.257 (−1.017)
Unrealistic targets (Dummy)	−0.242 (−1.354)		−0.352* (−1.740)

continued on page 168

Intervention	(1) Achievement of Project Objectives	(2) Achievement of Project Objectives	(3) Achievement of Project Objectives
Inadeq. M&E Framework (Dummy)	−0.228 (−1.396)		−0.281 (−1.602)
Overly complex design (Dummy)	−0.277 (−1.568)		−0.283 (−1.447)
Inadeq. stakeholder involvement (Dummy)		−0.358 (−1.191)	−0.0866 (−0.146)
Inadeq. borrower performance (Dummy)		−0.367** (−2.369)	−0.370* (−1.901)
Crisis /natural disaster (Dummy)		0.0640 (0.294)	0.0105 (0.0406)
Implement. delays (Dummy)		−0.0873 (−0.561)	−0.221 (−1.174)
Control variables	Yes	Yes	Yes
Observations	61	64	61

SOURCE: IEG.
* = 0.10 percent, ** = 0.05 percent, and *** = 0.01 percent significant.

Can good supervision overcome bad borrower performance, and vice versa? Can the developmental objectives be achieved when good supervision and complex design are present in the same project? As seen previously, many implementation problems are present in both successful and unsuccessful projects. Hence, to disentangle these combined effects a number of probit models with interaction terms is estimated. In earlier results, IEG observed that borrower performance appears to be the most important factor in achievement of project objectives, above Bank supervision. Does this result hold even when good borrower performance happens with inadequate supervision? Regression 1 of Table F.3 shows that even when paired with inadequate supervision, good borrower performance is associated with a positive achievement of the development objectives. The same, however, is not true in the opposite circumstance. In projects with good supervision and inadequate borrower performance the interaction term is significant and of the expected sign, indicating that good supervision can compensate for poor borrower performance, but the intercept has an unexpected negative sign, indicating that good supervision has a negative impact on project objectives (regression 2). These results seem to indicate that borrower performance is more important than supervision because results are more consistent.

Good borrower performance has also a positive effect when a crisis occurs. In these cases, in fact, the combined effect appears to be positive, indicating the ability of the borrower to overcome the crisis and achieve the developmental objectives. This is, however, not the case with delays. When these occur, not even a good borrower performance can ensure the achievement of developmental objectives (Table F.3, regressions 3 and 4).

Project supervision and design appear to work together. On one hand, if a project has not been properly designed, even good supervision will not be able to ensure the achievement of the project objectives. On the other hand, a good design in not enough to ensure the achievement of the project objectives when the project is affected by poor supervision (Table F.4, regressions 1 and 2). Finally, good targeting and a proper M&E system also work together. If either one is not present, the achievement of the development objective is put in doubt or not achieved.

TABLE F.3 Probit Interaction Term Regressions for World Bank Projects on Innovation and Entrepreneurship

	(1)	(2)	(3)	(4)
	Achievement of Project Objectives	Achievement of Project Objectives	Achievement of Project Objectives	Achievement of Project Objectives
Good Borrower Performance (Dummy)	0.176 (0.817)		0.338** (2.040)	0.402** (2.173)
Inadeq. supervision (Dummy)	−0.588* (−1.762)			
[Good Borrower * Inadeq. Supervision]	0.351** (2.119)			
Inadeq. stakeholder Involvement (Dummy)	0.00918 (0.0154)	0.00918 (0.0154)	−0.334 (−1.085)	−0.358 (−1.178)
Unrealistic targets (Dummy)	−0.392* (−1.784)	−0.392* (−1.784)		
Crisis /natural disaster (Dummy)	−0.106 (−0.387)	−0.106 (−0.387)		0.0765 (0.359)
Implement. delays (Dummy)	−0.200 (−0.988)	−0.200 (−0.988)	−0.0729 (−0.460)	
Inadeq. risk asses. (Dummy)	0.155 (0.515)	0.155 (0.515)		
Inadeq. technical design (Dummy)	−0.336 (−1.426)	−0.336 (−1.426)		
Inadeq. M&E framework (Dummy)	−0.303* (−1.705)	−0.303* (−1.705)		

	(1)	(2)	(3)	(4)
	Achievement of Project Objectives	Achievement of Project Objectives	Achievement of Project Objectives	Achievement of Project Objectives
Overly complex design (Dummy)	−0.411* (−1.881)	−0.411* (−1.881)		
Good supervision (Dummy)		−0.386 (−1.394)		
Inadeq. Borrower Performance (Dummy)		−0.936*** (−2.623)		
[Good superv.* Inadeq. Borrower perf.]		0.543** (2.119)		
Crisis /natural disaster (Dummy)			−0.0377 (−0.124)	
[Good Borr. Perf.* Crisis]			0.219 (0.617)	
Implement. delays (Dummy)				−0.0420 (−0.164)
[Good Borr. Perf.* Implement. Delays]				−0.0808 (−0.247)
Control variables	Yes	Yes	Yes	Yes
Observations	61	61	64	64

SOURCE: IEG.
* = 0.10 percent, ** = 0.05 percent, and *** = 0.01 percent significant.

TABLE F.4 Probit Interaction Term Regressions for World Bank Projects on Innovation and Entrepreneurship

	(1) Achievement of Project Objectives	(2) Achievement of Project Objectives	(3) Achievement of Project Objectives	(4) Achievement of Project Objectives
Good supervision (Dummy)	0.666** (2.456)			
Overly complex design (Dummy)	0.419 (1.344)			
[Good Superv* Compl. Design]	−0.820** (−2.230)			
Inadeq. risk asses. (Dummy)	0.227 (0.809)	0.227 (0.809)	0.134 (0.478)	0.134 (0.478)
Inadeq. technical design (Dummy)	−0.173 (−0.929)	−0.173 (−0.929)	−0.187 (−1.014)	−0.187 (−1.014)
Unrealistic targets (Dummy)	−0.327* (−1.655)	−0.327* (−1.655)		
Inadeq. M&E framework (Dummy)	−0.212 (−1.273)	−0.212 (−1.273)		
Not complex design (Dummy)		0.480** (2.232)		
Inadeq. Supervision (Dummy)		0.247 (0.894)		
[Not compl. Design* Inad. Supervision]		−0.761** (−2.230)		

	(1)	(2)	(3)	(4)
	Achievement of Project Objectives	Achievement of Project Objectives	Achievement of Project Objectives	Achievement of Project Objectives
Inadeq. Supervision (Dummy)			−0.274 (−1.220)	−0.274 (−1.220)
Good targeting (Dummy)			0.809** (2.408)	
Inadeq. M&E framework (Dummy)			0.350 (1.110)	
[Good targeting* Inadeq. M&E]			−0.792** (−2.129)	
Overly complex design (Dummy)			−0.287 (−1.499)	−0.287 (−1.499)
Good M&E (Dummy)				0.478** (2.213)
Unrealistic targets (Dummy)				−0.0288 (−0.142)
[Good M&E* Unrealistic targ.]				−0.740** (−2.129)
Control variables	Yes	Yes	Yes	Yes
Observations	61	61	61	61

SOURCE: IEG.
* = 0.10 percent, ** = 0.05 percent, and *** = 0.01 percent significant.

The results presented above are robust to the nonlinearity characteristic of the model. We tested our results by applying the method presented by Ai and Norton (2003). Table F.5 reports the values of the interaction terms throughout the whole sample. It demonstrates that all interaction terms maintain the same sign as in the above reported results.

In conclusion, the most important factors associated with the achievement of developmental objectives in World Bank projects are borrower performance, followed by quality of supervision and design, M&E, and appropriate targeting.

For IFC's investment projects in the portfolio, IEG looked first at detailed assessment of adequacy of at-entry assessment and work quality rating. In IFC projects with innovation and entrepreneurship elements, three implementation problems account for two-thirds of all problems: inadequate risk assessment, inadequate market assessment, and inadequate

TABLE F.5 World Bank Estimated Values of Interaction Terms Following Ai and Norton Method

Probit	Interaction Term	Mean	Median	Min	Max	Standard Deviation
3.1	Good borrower*inadequate supervision	0.601	0.687	0.092	0.753	0.191
3.2	Good supervision*inadeq. borrower perf.	0.601	0.687	0.092	0.753	0.191
3.3	Good borrower perf*crisis	0.184	0.193	0.088	0.231	0.040
3.4	Good borrower perf*implement. delays	−0.066	−0.071	−0.100	−0.005	0.026
4.1	Good supervi*compl. design	−0.761	−0.830	−0.995	−0.116	0.223
4.2	Not compl. design*inadeq. M&E	−0.761	−0.830	−0.995	−0.116	0.223
4.3	Good targeting*inadeq. M&E	−0.625	−0.711	−0.781	−0.141	0.167
4.4	Good M&E*unrealistic targ	−0.625	−0.711	−0.781	−0.141	0.167

SOURCE: IEG.
NOTE: M&E = monitoring and evaluation.

Distribution of Implementation Problems in IFC Investment Projects on Innovation and Entrepreneurship

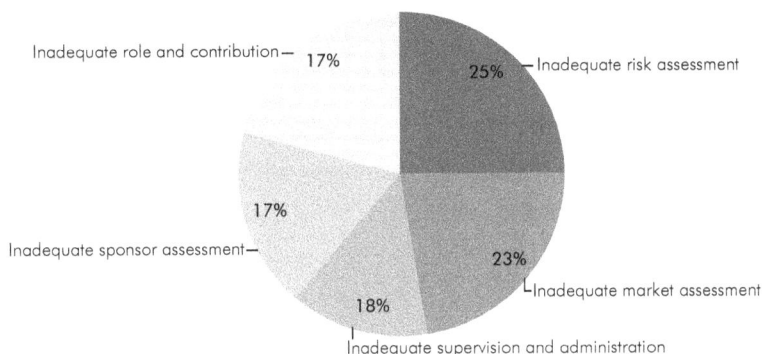

SOURCE: IEG.

supervision (Figure F.2). However, because—as in the World Bank portfolio—implementation setbacks occur in both IFC projects that achieve their developmental objectives and those that do not, IEG attempts to understand which setback is more important within a multivariate regression framework (Table F.6). Again, IEG controls for fixed factors such as project type, sponsor risk, and market risk, along with sector, region, amount of IFC lending, lessons learned, and country gross domestic product growth.

TABLE F.6 Problems in IFC Investment Projects on Innovation and Entrepreneurship

Problem	Developmental Objective	
	Not Achieved	Achieved
Inadequate sponsor assessment	42	5
Inadequate market assessment	54	10
Inadequate risk assessment	58	12
Inadequate supervision and administration	36	14
Inadequate role and contribution	40	7
	230	48

SOURCE: IEG.

Inadequate Skill Mix: The Bank team did not include needed expertise during the design or supervision. For example, the Bangladesh Agricultural Services Innovation and Reform supervision team did not have a continuous presence of competent agribusiness expertise. Presence of such expertise would have helped address the commercial and marketing issues.

Inadequate Risk Assessment: The factors that could potentially affect the project were not identified at the project design stage. For example, the Croatia Farmer Support Services Project did not identify two risks, which ultimately constrained project implementation: political lethargy in the economic reform process and the loss of "civicness" in the post-war social environment that seriously affected farmer willingness to associate.

Inadequate Technical Design: The project did not have a clear link between its inputs and outcomes and paid inadequate attention to realities on the ground, the choice of the instruments, political or institutional analysis, prior analytic work that lead to issues in the implementation of the project. For example, the Tunisia Agriculture Support Project paid insufficient attention to existing private sector farmer-to-market value chains and made unrealistic assumptions about the pace of development of representative producer organizations. The Uganda Agriculture Project was not specific and did not include measures to institutionalize and strengthen interagency cooperation (technical weakness). In the Brazil Development Policy Loan, the Bank's misreading of the changing circumstances led to a long delay in the effectiveness.

Inadequate Supervision: The Bank team was not timely or proactive to identify and take actions to correct deficiencies/issues during the implementation (ICR Review rating).

Ambitious Design or Unrealistic Targets: The project scale, activities, duration, and cost was ambitious.

Overly Complex Design: The project included many activities that hindered the management, supervision, and so forth. For example, the Bosnia Small-Scale Commercial Agriculture Development Project's design was complex, involving many overlapping components and activities, particularly in the market integration component, that made it difficult for the project units in each entity to manage and focus their activities, at least at the outset. This required changes in the structure and design of the project management based on more simple subject matter approaches.

Inadequate M&E framework, Poor Data Quality/Indicators: The project did not have well-designed M&E and there were issues in the implementation of and use of M&E.

Inadequate Stakeholder Involvement: The project did not consult relevant stakeholders adequately. For example, the existing weaknesses of HORTEX could have been addressed better had there been more interactions with the existing network of Bangladesh Fruit, Vegetable and Allied Products Exporters Association, which has more than 35 active members.

Implementation Disrupted by a Crisis/Natural Disaster: The project was disrupted by a financial or political crisis or natural disaster.

Inadequate Borrower Performance: The rating is based on several dimensions, including government and implementing agencies' commitment, ownership, capacity, and coordination (ICR Review rating).

After controlling for fixed factors, IEG observes that adequate ex ante sponsorship assessment and adequate ex ante market assessment have a strong and positive impact on the probability of achieving the developmental objective. Each increases the chances of achieving it by approximately 20–30 percent. Lessons learned also show a positive and significant impact, even though the sample size is much smaller when this control variable is included (Table F.7). The impact of IFC lending is significant, but marginal in size.

The IFC work quality ratings from XPSRs are then included in this basic probit regression. The results show that quality of supervision, role and contribution, and sponsor assessment have a positive and significant impact by raising the probability of achieving the developmental objective by approximately 30 percent (Table F.7, regression 1).

As observed earlier with the World Bank portfolio, project problems occur both when the developmental objectives are achieved and when they are not. Hence, in an attempt to disentangle the effects of all these variables and establish which ones are more important, IEG uses interaction terms in the probit model. This analysis shows that only market assessment and supervision are significant factors in the interaction. More specifically, inadequate market assessment has a negative impact on the probability of achieving the developmental objective, even when projects have good supervision (Table F.8, regression 2). In contrast,

TABLE F.7 Probit Regression for IFC Investment Projects on Innovation and Entrepreneurship

	(1) Achievement of Project Objectives	(2) Achievement of Project Objectives	(3) Achievement of Project Objectives
Project type risk (dummy)	0.0289 (0.364)	−0.208 (−1.427)	0.0370 (0.446)
Ex ante sponsor risk (dummy)	−0.272*** (−3.442)	−0.221 (−1.529)	−0.294*** (−3.606)
Ex ante market risk (dummy)	−0.253*** (−3.023)	−0.303** (−1.994)	−0.222** (−2.528)
GDP growth	0.00761 (1.068)	−0.0208 (−1.147)	0.00937 (1.299)
Lessons learned (dummy)		0.390** (2.508)	
IFC lending $(log)			4.52e−06** (1.980)
Observations	181	53	174

SOURCE: IEG.
* = 0.10 percent, ** = 0.05 percent, *** = 0.01 percent significant.

bad supervision has a negative effect, but adequate market assessment compensates for such effect, and the combined effect is almost zero (Table F.7, regression 3).

The results are robust to the nonlinearity characteristic of the model (Ai and Norton, 2003). Table F.9 shows that the interaction terms maintain the same signs as in the original model.

Probit Interaction Term Regressions for IFC Investment Projects on Innovation and Entrepreneurship

	(1)	(2)	(3)
	Achievement of Project Objectives	**Achievement of Project Objectives**	**Achievement of Project Objectives**
Inadeq. Sponsor assessment (Dummy)	−0.372** (−2.070)	−0.400** (−2.114)	−0.400** (−2.114)
Inadeq. Role and Contribution (Dummy)	−0.285** (−2.029)	−0.327** (−2.303)	−0.327** (−2.303)
Inadeq. Risk assessment (Dummy)	0.0166 (0.0526)	−0.0219 (−0.0639)	−0.0219 (−0.0639)
Inadeq. Market assessment (Dummy)	−0.420 (−1.489)		
Inadeq. Supervision and admin. (Dummy)	−0.278** (−2.038)		
Inadeq. Market assess. (Dummy)		−0.701** (−2.177)	
Good Supervision		0.0164 (0.0783)	
[Inad. Market assess.* Good Supervision]		0.496* (1.876)	
Adequate Market assessment (Dummy)			0.252 (0.785)
Inadequate Supervision (Dummy)			−0.532** (−2.534)

continued on page 180

	(1)	(2)	(3)
	Achievement of Project Objectives	Achievement of Project Objectives	Achievement of Project Objectives
[Adeq. Market assess.* Inadeq. Supervision]			0.457*
			(1.876)
Control variables	Yes	Yes	Yes
Observations	171	171	171

SOURCE: IEG.
* = 0.10 percent, ** = 0.05 percent, and *** = 0.01 percent significant.

TABLE F.9 IFC Estimated Values of Interaction Terms Following Ai and Norton Method

Probit	Interaction Term	Mean	Median	Min.	Max.	Standard Deviation
7.2	Inad. Market assess. * good supervision	0.369	0.466	0.007	0.556	0.197
7.3	Adeq. Market assess.*inadeq. Supervision	0.369	0.466	0.007	0.556	0.197

SOURCE: IEG.

Endnote

[1] We follow the approach of Denizer, Kaufmann, and Kraay (2011).

References

Ai, Chunrong, and Edward C. Norton. 2003. "Interaction Terms in Logit and Probit Models." *Economic Letters*, 80:123–29.

Denizer, C., Daniel Kaufmann, and Aart Kraay. 2011. "Good Countries or Good Projects? Macro and Micro Correlates of World Bank Project Performance." Policy Research Working Paper 5646, World Bank, Washington, DC.

Appendix G
Knowledge Sharing

FIGURE G.1 Knowledge Shared Formally, by Type of Experience

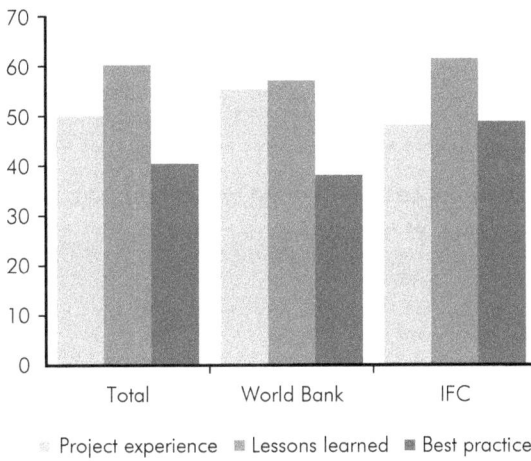

SOURCE: IEG.

FIGURE G.2 Distribution of Lessons Shared Formally, by Delivery Mechanism

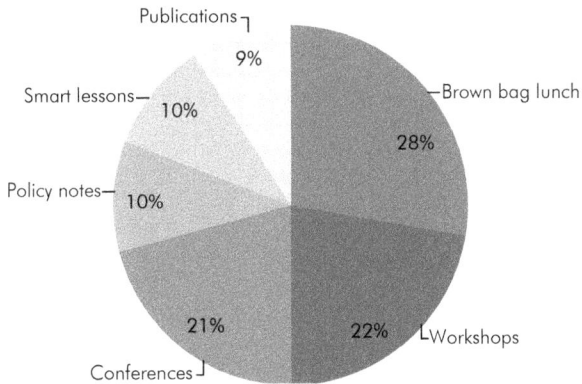

SOURCE: IEG.

FIGURE G.3 Distribution of Bank Group Counterpart Lesson-Sharing, by Level of Formality

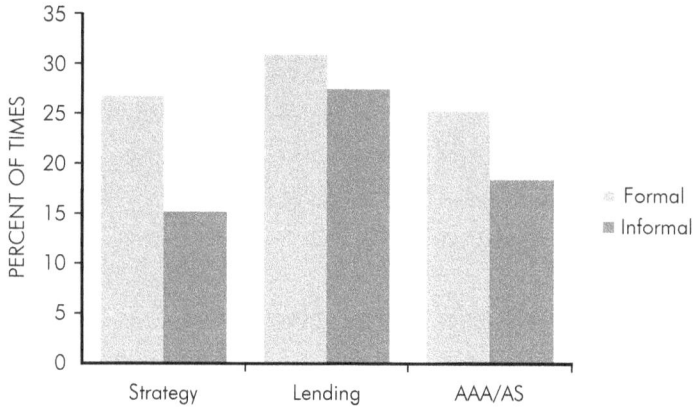

SOURCE: IEG.
NOTE: AAA = analytic and advisory activity; AS = Advisory Services.

FIGURE G.4 Overall Distribution of Lesson-Sharing within and across Networks, Regions, and Institutions, by Level of Formality

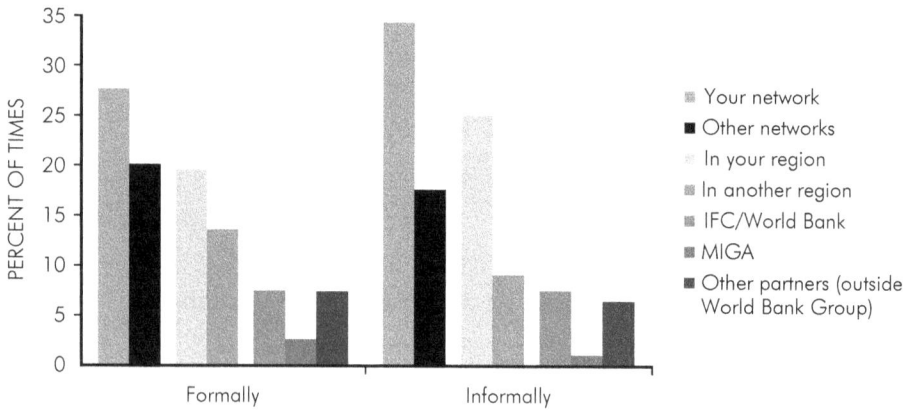

SOURCE: IEG.

Bibliography

Campos, Francisco, Aidan Coville, Ana M. Fernandes, Markus Goldstein, and David McKenzie. 2012. "Learning from the Experiments That Never Happened: Lessons from Trying to Conduct Randomized Evaluations of Matching Grant Programs in Africa." World Bank Policy Research Working Paper 6296, Washington, DC.

IEG (Independent Evaluation Group). 2009. *Improving Effectiveness and Outcomes for the Poor in Health, Nutrition, and Population: An Evaluation of World Bank Group Support since 1997.* Washington, DC: World Bank.

Jaumotte, F., and N. Pain. 2005. "Innovation Policies and Innovation in the Business Sector." Economics Department Working Paper 459, OECD, Paris.

World Bank. 2009a. *Education Sector Strategy.* Washington, DC: World Bank.

——. 2009b. *World Bank Group Agriculture Action Plan: FY 2010–2012.* Washington, DC: World Bank.

——. 2005. *Education Sector Strategy Update.* Washington, DC: World Bank.

——. 2003. *Reaching the Rural Poor: A Renewed Strategy for Rural Development.* Washington, DC: World Bank.

——. 2002. *Private Sector Development Strategy—Directions for the World Bank Group.* Washington, DC: World Bank.

——. 1999. *Education Sector Strategy Update.* Washington, DC: World Bank.

www.ingramcontent.com/pod-product-compliance
Lightning Source LLC
Chambersburg PA
CBHW080538220326
41599CB00032B/6307